Revising Oral Pharmacokinetics, Bioavailability and Bioequivalence Based on the Finite Absorption Time Concept

Panos Macheras • Athanasios A. Tsekouras

Revising Oral Pharmacokinetics, Bioavailability and Bioequivalence Based on the Finite Absorption Time Concept

Springer

Panos Macheras
Faculty of Pharmacy
National and Kapodistrian
University of Athens & ATHENA
Research Center
Athens, Greece

Athanasios A. Tsekouras
Department of Chemistry
National and Kapodistrian
University of Athens & ATHENA
Research Center
Athens, Greece

ISBN 978-3-031-20024-3　　ISBN 978-3-031-20025-0 (eBook)
https://doi.org/10.1007/978-3-031-20025-0

This Springer imprint is published by the registered company Springer Nature Switzerland AG
The registered company address is: Gewerbestrasse 11, 6330 Cham, Switzerland

Dedicated to our children and grandchildren
Evangelos, Spiros, Andreas, Giorgos,
Artemis, Thalis, Stefanos, Panos and Amelia

Foreword

It is rare and refreshing to see a well-researched treatise that challenges a long-held scientific approach that obviously deserves scrutiny. Real-world oral pharmacokinetic curves not uncommonly show unexplained up-and-down patterns in the ascending or absorption part of overall bell-shaped concentration-time trajectories, followed by a sharp decline in concentration after the observed maximum concentration (C_{max}). Such datasets are not well modeled using conventional compartmental models. The long overlooked Finite Absorption Time (FAT) concept, as manifested in the Physiologically Based Finite Time Pharmacokinetic (PBFTPK) models described in this book, offers a sound theoretical foundation to address such shortcomings of conventional oral pharmacokinetic models. It may come as a delight to pharmaco metricians to be able to adequately fit oral pharmacokinetic curves with strange double peaks or zigzag patterns in the absorption portion of the curve, often attributed to randomness of data or experimental aberrations using conventional approaches.

As discussed in Chap. 7, the FAT/PBFTPK approach may contribute a supportive basis for regulatory recommendations to using partial AUC values (pAUC) for bioequivalence (BE) assessments. pAUC values can be usefully employed when a formulation change leads to a modified exposure response relationship without affecting C_{max} and AUC. For example, the FDA recommended the use of partial AUC pAUC determinations for demonstrating BE of generic zolpidem extended-release tablets and methylphenidate hydrochloride ER capsules in 2011. As of June 2022, the FDA has issued 44 product-specific guidances recommending the use of pAUC to determine BE for drugs submitted via an ANDA. Consideration for using pAUCs in BE metrics and the selection of time intervals to truncate the AUC are both drug- and formulation-specific. PBFTPK modeling can guide understanding of how a formulation interacts with segmental gastrointestinal physiology as reflected in the observed PK curves.

Critical evaluation is required of any new scientific approach that is proposed to improve on a long-practiced basis for characterizing PK and BE, which are key to evaluation of new or generic drugs. The authors are to be commended for challenging oral pharmacokinetic traditional models of drug absorption that perform poorly in some circumstances. Via this well-written book, the scientific community can learn of the FAT concept and evaluate its value proposition seriously.

Disclaimer: This article reflects the views of the authors and should not be construed to represent FDA's views or policies.

NDA Partners, a ProPharma Group Company, Washington, DC, USA

Department of Bioengineering and Therapeutic Sciences, University of California at San Francisco, CA, USA

Carl Peck

Division of Quantitative Methods and Modeling, Office of Research and Standards, US Food and Drug Administration, Silver Spring, MD, USA

Liang Zhao

Preface

According to the Editor of Journal of Pharmacokinetics and Pharmacodynamics Dr Peter Bonate [1] "in a few short years, quantitative systems pharmacology (QSP) has become a major tool available to pharmacometricians to improve decision making in drug development, so much so that today pharmacometrics can be broadly classified into three groups: population-based methods, physiologically-based pharmacokinetics (PBPK), and Quantitative Systems Pharmacology (QSP). Recently, we are starting to see the emergence of a fourth field: machine learning." This is so since science progresses. It evolves. New knowledge is created. Pharmacometrics is no exception [1].

In the midst of these dramatic changes, we realized in 2019 that a wrong assumption that breaks oral pharmacokinetics was used and extensively applied since 1953 [2]. In fact, the infinite time of oral drug absorption was conceived from the first day of the birth of pharmacokinetics when F. H. Dost introduced the term pharmacokinetics [2]. He adopted the function developed by H. Bateman [3] back in 1910 for the decay of the radioactive isotopes to describe oral drug absorption as a first-order process. We unveiled this false hypothesis relying on common wisdom, i.e., drugs are absorbed in finite time. This false assumption had dramatic effects on the evolution of oral pharmacokinetics, but most importantly on the bioavailability and bioequivalence concepts and metrics. Accordingly, the title of this book could be "*Unveiling the wrong assumption that breaks oral pharmacokinetics: Drugs are absorbed in finite time.*" Instead, we utilize a different title which places emphasis on the "*revision*" of the three major disciplines of biopharmaceutics-pharmacokinetics, namely, oral pharmacokinetics, bioavailability, and bioequivalence under the prism of Finite Absorption Time (FAT).

In oral pharmacokinetics, the absorption rate constant became the sole parameter for expressing quantitatively the rate of drug absorption in classical and population pharmacokinetic studies. However, it was found to be the most variable parameter with non-physiological meaning having units ($time^{-1}$), not allowing a valid interspecies or pediatric scaling and relying on the unphysical assumption of infinite time of absorption [4]. Twenty years ago or so when the development of PBPK models started, the assessment of the rate of drug input was based on permeability estimates, namely, the PBPK models abandon the use of absorption rate constant for the assessment of the drug's input rate. Our work was based on the physiologically sound FAT concept [5]; thus, the relevant Physiologically Based Finite Time Pharmacokinetic (PBFTPK) models developed were found to be a powerful tool for the pharmacokinetic analysis of oral concentration, time data. The software developed is made available to the readers as supplementary electronic material (see http://sn.pub/PtsM9h).

In bioavailability studies, the area under the curve from zero to infinity $[\mathrm{AUC}]_0^\infty$, which corresponds to the indefinite integral of the concentration-time function describing the time course of drug in the body, wrongly became the golden metric for the extent of drug absorption. In reality, $[\mathrm{AUC}]_0^\infty$ is an ideal exposure metric; intuitively, the $[\mathrm{AUC}]_0^\tau$, where τ is the FAT, is the proper metric for the extent of drug absorption. In parallel, $C_{\max}$ is currently used as an absorption rate metric. Under the FAT concept, the concentration at time τ, $C(\tau)$ corresponds to drug concentration at the termination of the drug absorption process(es). In this vein, the numerical value of the observed maximum blood drug concentration equal to or greater than $C(\tau)$ should be used as such. This means that the magnitude of its difference between reference and test formulations in bioequivalence studies should be specified on pharmacological-pharmacodynamic basis for each one of the drugs examined. For example, critical dose drugs with narrow therapeutic index, e.g., cyclosporine, can have a smaller absolute difference and/or an upper/lower boundary for the test and reference formulations. These considerations point to the abolishment of the term "rate" in the definition of bioavailability and the use of the relevant parameter $C_{\max}$ accompanied with statistical criteria as an indicator of the rate of absorption.

The book is divided into two parts. In Part I, the first two chapters are devoted to the mathematics associated with the unphysical hypothesis of infinite absorption time as well as the extensive use of the absorption rate constant in biopharmaceutics and pharmacokinetics since 1953. Chapter 3 focuses on the development of the FAT concept, while the relevant PBFTPK models are described in Chap. 4. In Part II, Chap. 5 relies on the historical aspects of the bioavailability and bioequivalence concepts. The evolution of bioequivalence studies for the establishment of therapeutic equivalence of test and reference formulations is the subject of Chap. 6. Bioavailability is analyzed under the prism of the FAT concept in Chap. 7. Methodologies for the estimation of absolute bioavailability from oral data exclusively are reported for the first time. In Chap. 8, a methodology towards the revision of the bioequivalence assessment is presented.

This book is intended for academics/students or scientists working in pharmaceutical industries, regulatory agencies, and contract research organizations. It can be used for teaching purposes in undergraduate courses dealing with biopharmaceutics, pharmacokinetics, and biomedical engineering. However, the content of Chap. 4 and the relevant PBFTPK software applications are suitable for postgraduate courses of these disciplines. In parallel, as already mentioned, the use of the PBFTPK software is made possible to the readers through the electronic supplementary material accompanying this book.

A number of obvious applications of the FAT concept for pharmacometricians to important topics have not been included in this edition. For example, all software dealing with oral drug absorption in PBPK modeling work generates percent absorbed versus time profiles which are distorted and shifted to the right. This is the net result of the integral from zero to infinity applied to solve the differential equations describing the rate of the drug input processes. In a similar vein, in vitro in vivo correlations require reconsideration since all methodologies applied, c.f., Wagner-Nelson, Loo-Riegelman, and deconvolution techniques for the %absorbed versus time curve, utilize $[AUC]_0^\infty$ and not $[AUC]_0^\tau$ in the numerical integration step. In oral, pulmonary and intranasal pharmacokinetic and pharmacokinetic–pharmacodynamic population studies, the structural models used so far do not include the duration of absorption, τ as a fundamental parameter of the model. This is particularly so for studies under fed conditions; the PBFTPK models developed herein are the most suitable to be used in population Pharmacokinetic–Pharmacodynamic studies. Similar applications to interspecies and pediatric scaling can be also envisaged.

The book was conceived the summer of 2021 when we realized that the estimation of absolute bioavailability can be achieved from oral data exclusively.

P.M. expresses his gratitude to the current Minister of Education of Greece Mrs Niki Kerameos for implementing (Law 4957/2022) his plea to allow professors after obligatory retirement to mentor graduate students, which was included in the acknowledgement section of his recent publications [4–9].

References

1. Bonate PL ((2022)) Editor's note on the themed issue: integration of machine learning and quantitative systems pharmacology. J. Pharmacokin. Pharmacodyn 49:1–3 https://doi.org/10.1007/s10928-022-09803-1
2. Dost FH (1953) Der Blutspiegel. Kinetik der Konzentrationsabläufe in der Kreislaufflüssigkeit. Thieme, Leipzig
3. Bateman H (1910) The solution of a system of differential equations occurring in the theory of radioactive transformations. Proc Cambridge Philos Soc Math Phys Sci 15:423–427. https://www.biodiversitylibrary.org/item/97262#page/487/mode/1up or https://archive.org/details/cbarchive_122715_solutionofasystemofdifferentia1843
4. Macheras P (2019) On an unphysical hypothesis of Bateman equation and its implications for pharmacokinetics. Pharm Res 36:94. https://doi.org/10.1007/s11095-019-2633-4
5. Macheras P, Chryssafidis P (2020) Revising pharmacokinetics of oral drug absorption: I models based on biopharmaceutical/physiological and finite absorption time concepts. Pharm Res 37:187. https://doi.org/10.1007/s11095-020-02894-w

6. Macheras P, Kosmidis K, Chryssafidis P (2020) Demystifying the spreading of pandemics I: The fractal kinetics SI model quantifies the dynamics of COVID-19, meRxiv. https://doi.org/10.1101/2020.11.15.20232132
7. Chryssafidis P, Tsekouras AA, Macheras P (2021) Revising pharmacokinetics of oral drug absorption: II Bioavailability-bioequivalence considerations. Pharm Res 38:1345–1356. https://doi.org/10.1007/s11095-021-03078-w
8. Tsekouras AA, Macheras P (2021) Re-examining digoxin bioavailability after half a century: Time for changes in the bioavailability concepts. Pharm Res 38: 1635–1638. https://doi.org/10.1007/s11095-021-03121-w
9. Chryssafidis P, Tsekouras AA, Macheras P (2022) Re-writing oral pharmacokinetics using physiologically based finite time pharmacokinetic (PBFTPK) models. Pharm Res 39:691–701. https://doi.org/10.1007/s11095-022-03230-0

NOTE: The Electronic Supplementary Material of this book can be accessed at http://sn.pub/PtsM9h or via this QR.

Athens, Greece
August 2022

Panos Macheras
Athanasios A. Tsekouras

Contents

Part I
Oral Pharmacokinetics

Chapter 1
From Bateman (1910) to Dost (1953)

Time is what we want most, but we use worst.
William Penn (1644–1718)

Contents

Abstract This chapter focuses on one of the first applications of first-order kinetics in Physics. It introduces the Bateman equation based on a model with three radioactive species, namely, mother, daughter, and granddaughter. This equation was adopted by Dost in 1953 to describe the concentration–time curve of drug absorption after oral administration assuming one-compartment model system.

Keywords Bateman equation · First-order kinetics · Bioavailability parameters · AUC · C_{max}

P. Macheras, A. A. Tsekouras, *Revising Oral Pharmacokinetics, Bioavailability and Bioequivalence Based on the Finite Absorption Time Concept*,
https://doi.org/10.1007/978-3-031-20025-0_1

1.1 Harry Bateman: First-Order Kinetics in Physics (1910)

According to the bibliographical memoir [1] of F. Murnaghan, Harry Bateman was born in Manchester, England, on May 29, 1882. Bateman attended Manchester Grammar School where he first grew to love mathematics, and in his final year he won a scholarship to Trinity College, Cambridge. He went up to Cambridge in 1900 and was Senior Wrangler in the Mathematical Tripos examinations of 1903 when he was awarded his B.A. He became a Reader in mathematical physics at the University of Manchester in 1907. Bateman immigrated to the United States in 1910. He received many honors for his contributions, including election to the Royal Society of London in 1928 and election to the National Academy of Sciences in Washington in 1930. He died in Pasadena, California, on January 21, 1946.

According to his biographer, Bateman solved systems of differential equations [2] discovered by Rutherford [3] which describe radioactive decay. In nuclear physics, the *Bateman equation* is a mathematical model describing abundances and activities in a decay chain as a function of time, based on the decay rates and initial abundances. The model was formulated by Ernest Rutherford in 1905 [3] and the analytical solution was provided by Harry Bateman in 1910 [2]. For the simple case of a chain of three isotopes (mother, daughter, and granddaughter), Fig. 1.1, the corresponding Bateman equation reduces to an equation with two exponentials for the abundance of the daughter species N_{daugther}:

$$N_{\text{daughter}} = \frac{\lambda_{\text{m}} N_{\text{m0}}}{\lambda_{\text{m}} - \lambda_{\text{d}}} \left(e^{-\lambda_{\text{d}} t} - e^{-\lambda_{\text{m}} t} \right) \tag{1.1}$$

where λ_{m} and λ_{d} are first-order rate constants for the transition of the mother to daughter species and the transition from the daughter to granddaughter species, respectively; N_{m0} is the initial abundance of the mother species.

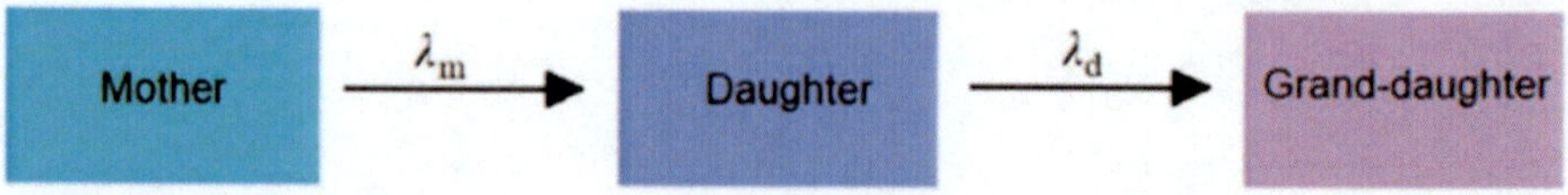

Fig. 1.1 The model of a chain of three isotopes. The transition rates are only allowed from one species to the next, but never in the reverse sense

1.2 Friedrich Hartmut Dost: First-Order Kinetics in Pharmacokinetics (1953)

Dost (1910–1985) introduced the term pharmacokinetics in 1953 in his text *Der Blutspiegel. Kinetic der Konzentrationsabläufe in der Kreislaufflüssigkeit* [4]. This outstanding book for its time fully covered the so-called one-compartment open model with its various forms of input [5]. We recall here Eq. (1.2) from Dost's book. Equation (1.2) describes the concentration of drug in blood *C*(*t*)[1] as a function of time for the linear one-compartment model with first-order absorption and elimination:

$$C(t) = \frac{\mathrm{FD}k_{\mathrm{a}}}{V_{\mathrm{d}}(k_{\mathrm{a}} - k_{\mathrm{el}})}\left(e^{-k_{\mathrm{el}}t} - e^{-k_{\mathrm{a}}t}\right) \tag{1.2}$$

where F is the bioavailable fraction of dose D, V_{d} is the volume of distribution, k_{a} is the first-order rate constant of absorption, and k_{el} is the elimination first-order rate constant. All fundamental parameters associated with bioavailability, namely, $[\mathrm{AUC}]_0^{\infty}$, C_{max} and t_{max} (see Eqs. 1.4–1.6, respectively) are derived from Eq. (1.2).

The similarity of Eqs. (1.1) and (1.2) is obvious. In fact, Dost replaced the abundance of the daughter species with the concentration of drug in blood, Fig. 1.2.

According to Eq. (1.2), both absorption and elimination processes are first-order processes and run concurrently from zero to infinity. Indeed, thousands of experimental observations have shown that the decay of isotopes follows first-order kinetics. Similarly, the prevailing first-order character of the elimination rate of drugs has been verified in numerous pharmacokinetic studies. In this context, everyone can recall the concept of half-life, $t_{1/2}$ of radioisotopes we became acquainted with in our high-school days. This was further extended to the half-life

[1] Throughout the book we are concerned with the concentration of a drug in the blood which will be represented with the letter *C* without a subscript. The concentration of the same drug in other parts of the body will not be considered in detail since we do not deal with physiological circulatory models.

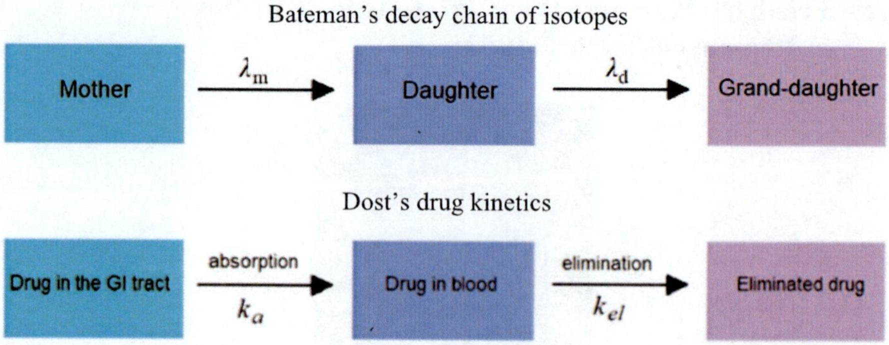

Fig. 1.2 Harry Bateman's vis-a-vis Friedrich Hartmut Dost's kinetic considerations

of drugs in pharmacokinetics and its relation with the elimination rate constant ($t_{½} = \ln 2/k_{el}$).

On the contrary, the infinite time of drug absorption is not physiologically sound since drugs are not absorbed beyond the absorptive sites in the gastrointestinal (GI) tract. In fact, oral drug absorption takes place in a certain period of time in accordance with the biopharmaceutical properties of the drug as well as the physiological gastric, intestinal, and colon transit times reported in the literature [6]. The development of Finite Absorption Time (FAT) concept relies on this discrepancy as delineated in Chap. 3. Accordingly, the Physiologically Based Finite Time Pharmacokinetic (PBFTPK) models, described in Chap. 4, are the core theme of this book.

1.3 Simulations Based on Bateman's Equation

Figure 1.3 shows a typical concentration–time plot based on Eq. (1.2) using various values of the parameters k_a and k_{el}. It can be seen that the higher the ratio k_a/k_{el} the steeper the increase of concentration during the initial phase of drug absorption.

Fig. 1.3 Concentration–time plots for various k_a and k_{el} values, Eq. (1.2)

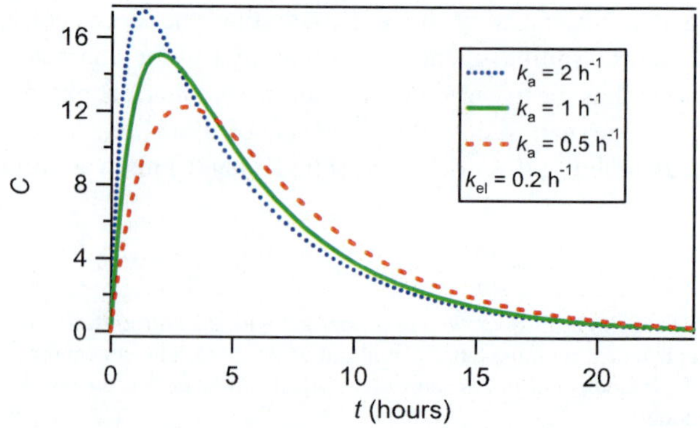

From the early days of the use of Eq. (1.2), three cases were considered, i.e., the classical, $k_a > k_{el}$, *the flip-flop,* $k_a < k_{el}$, and the special one when $k_a = k_{el} = k$. In the latter case, Eq. (1.3) replaces Eq. (1.2):

$$C(t) = \frac{\mathrm{FD}kt}{V_d} e^{-kt} \tag{1.3}$$

During the pre-computer era of pharmacokinetics, the problem of the relative magnitude of the rate constants was a hot topic [7–12].

1.4 The Fundamental Metrics of Bioavailability/Bioequivalence Studies

Equation (1.2) can be used as the starting point for the consideration of the bioavailability/ bioequivalence issues since the relevant parameters for the extent of absorption $[\mathrm{AUC}]_0^\infty$ and the rate of absorption C_{max} are derived upon integration of Eq. (1.2) and equating the first derivative of Eq. (1.2) with zero, respectively.

$$\int_0^\infty C(t)\mathrm{d}t = \int_0^\infty \frac{\mathrm{FD}k_a}{V_d(k_a - k_{el})}\left(e^{-k_{el}t} - e^{-k_a t}\right)\mathrm{d}t = [\mathrm{AUC}]_0^\infty = \frac{\mathrm{FD}}{\mathrm{CL}} \tag{1.4}$$

$$C_{max} = \frac{\mathrm{FD}}{V_d}\left(\frac{k_{el}}{k_a}\right)^{\frac{k_a}{k_a - k_{el}}} \tag{1.5}$$

$$t_{max} = \frac{1}{k_a - k_{el}} \ln\left(\frac{k_a}{k_{el}}\right) \tag{1.6}$$

where CL is the drug clearance ($\mathrm{CL} = k_{el}V_d$) and t_{max} is the time at which C_{max} is observed. According to Eq. (1.4) $[\mathrm{AUC}]_0^\infty$ is proportional to the fraction of bioavailable dose. Although this proportionality makes $[\mathrm{AUC}]_0^\infty$ the ideal metric for the extent of absorption in bioavailability/bioequivalence studies, a question is raised in Chap. 2. Bioavailability and bioequivalence are reconsidered in Chaps. 7 and 8 in the light of the finite absorption time concept. Finally, a graph depicting the three parameters is shown in Fig. 1.4.

All above allow us to recall Penn's quote in the subtitle of this chapter. In fact, the infinite time of drug absorption can be considered as the nemesis of oral pharmacokinetics, bioavailability, and bioequivalence studies.

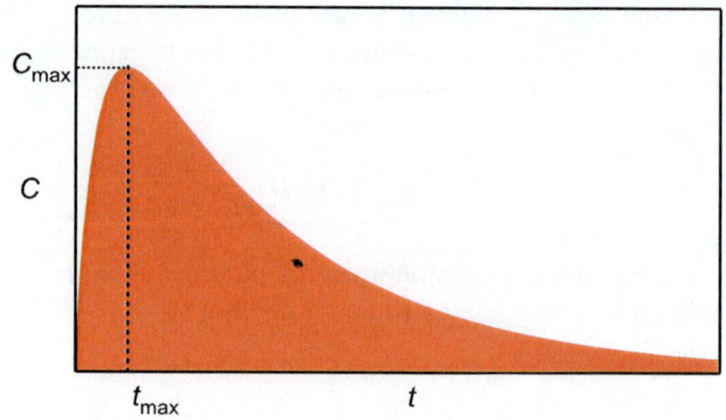

Fig. 1.4 A concentration–time curve generated from Eq. (1.2) showing bioavailability parameters. The shaded area depicts $[\mathrm{AUC}]_0^{\infty}$.

References

1. Murnaghan FD (1948) Harry Bateman 1882-1946. Bull Am Math Soc 54:88–103. https://doi.org/10.1090/S0002-9904-1948-08955-8
2. Bateman H (1910) The solution of a system of differential equations occurring in the theory of radioactive transformations. Proc Cambridge Philos Soc Math Phys Sci 15:423–427. https://www.biodiversitylibrary.org/item/97262#page/487/mode/1up or https://archive.org/details/cbarchive_122715_solutionofasystemofdifferentia1843
3. Rutherford E (1905) Radio-activity. University Press, Cambridge, p 331
4. Dost FH (1953) Der Blutspiegel. Kinetik der Konzentrationsabläufe in der Kreislaufflüssigkeit. Thieme, Leipzig
5. Wagner JG (1981) History of pharmacokinetics. Pharmacol Ther 12:537–562. https://doi.org/10.1016/0163-7258(81)90097-8
6. Abuhelwa A, Foster D, Upton R (2016) A quantitative review and meta-models of the variability and factors affecting oral drug absorption-Part II: gastrointestinal transit time. AAPS J 18:1322–1333. https://doi.org/10.1208/s12248-016-9953-7
7. Bialer M (1980) A simple method for determining whether absorption and elimination rate constants are equal in the one-compartment open model with first-order processes. J Pharmacokinet Biopharm 8:111–113. https://doi.org/10.1007/BF01059453
8. Patel IH (1984) Concentration ratio method to determine the rate constant for the special case when $k_a = k_e$. J Pharm Sci 73:859–861. https://doi.org/10.1002/jps.2600730648
9. Macheras P (1985) A graphical approach for determining whether absorption and elimination rate constant are equal in the one-compartment open model with first-order processes. J Pharm Sci 74:582–584. https://doi.org/10.1002/jps.2600740521
10. Macheras P (1985) Developments in the concentration ratio method. J Pharm Sci 74:1021. https://doi.org/10.1002/jps.2600740927
11. Macheras P (1987) Method of residuals: estimation of absorption and elimination rate constants having comparable values. Biopharm Drug Disp 8:47–56. https://doi.org/10.1002/bdd.2510080106
12. Macheras P (1987) Improvement without computer assistance of the graphically estimated parameters of the linear one-compartment model. Biopharm Drug Disp 8:387–394. https://doi.org/10.1002/bdd.2510080409

Chapter 2
The Unphysical Hypothesis of Infinite Absorption Time

Everything should be made as simple *as possible, but no* simpler.
Albert Einstein (1879–1955)

Contents

Abstract This chapter describes the extensive use of first-order kinetics in oral drug absorption phenomena along with the governing role of the absorption rate constant. It also includes the use of a time-dependent coefficient based on fractal kinetics principles as a more realistic way of describing drug gastrointestinal absorption. The last portion of the chapter focuses on the unphysical assumption associated with the use of Bateman equation in pharmacokinetics.

Keywords Bateman equation · Absorption rate constant · BCS · BDDCS · Fractal kinetics · FDA guideline · EMA guideline

P. Macheras, A. A. Tsekouras, *Revising Oral Pharmacokinetics, Bioavailability and Bioequivalence Based on the Finite Absorption Time Concept*,
https://doi.org/10.1007/978-3-031-20025-0_2

2.1 The Biopharmaceutical–Pharmacokinetic Basis of the Bateman Equation

Basically, drugs pass through the gastrointestinal membranes by passive diffusion. Fick's laws of diffusion describe the flux of solutes (drugs) undergoing classical diffusion. The simplest system to consider is a solution of a drug with two regions of different concentrations, C_{GI} at the absorption site of the gastrointestinal (GI) lumen and blood concentration, C, of a boundary (GI membrane) separating the two regions. The driving force for drug transfer is the concentration gradient between the concentrations of the drug molecules in the two regions. Thus, the rate of penetration can be written:

$$\text{Rate of Penetration} = P \cdot (\mathrm{SA}) \cdot (C_{\mathrm{GI}} - C) \tag{2.1}$$

where P is the permeability of drug expressed in velocity units (length/time) and SA is the surface area of the membrane in $(\text{length})^2$ units. The sheer size of the body, by diluting absorbed drug, tends to maintain sink conditions, in which C is much smaller than C_{GI}, therefore,

$$\text{Rate of Penetration} = P \cdot (\mathrm{SA}) \cdot (C_{\mathrm{GI}}) \tag{2.2}$$

Equation (2.2) can be written in terms of drug amount, A_{GI} assuming that the volume of fluid at the absorption site V_{GI} remains relatively constant,

$$\text{Rate of Penetration} = P \cdot (\mathrm{SA}) \cdot \frac{A_{\mathrm{GI}}}{V_{\mathrm{GI}}} = k_{\mathrm{a}} A_{\mathrm{GI}} \tag{2.3}$$

where k_{a} is the absorption rate constant expressed in $(\text{time})^{-1}$ units, which is equal to $P(\mathrm{SA})/V_{\mathrm{GI}}$.

In all pharmacokinetic textbooks, the classical analysis of one-compartment model starts from Eq. (2.4) assuming a first-order decrease of the amount of drug, A_{GI}, in agreement with Eq. (2.3):

$$\frac{\mathrm{d}A_{\mathrm{GI}}}{\mathrm{d}t} = -k_{\mathrm{a}} A_{\mathrm{GI}} \tag{2.4}$$

which upon integration from $t = 0$, $A_{\mathrm{GI}} = \mathrm{FD}$ to $t = t$, $A_{\mathrm{GI}} = A_{\mathrm{GI}}$ one obtains:

$$A_{\mathrm{GI}}(t) = \mathrm{FD} \cdot \mathrm{e}^{-k_{\mathrm{a}} t} \tag{2.5}$$

Equation (2.5) is further coupled with the differential equation describing the change of drug concentration in blood, C:

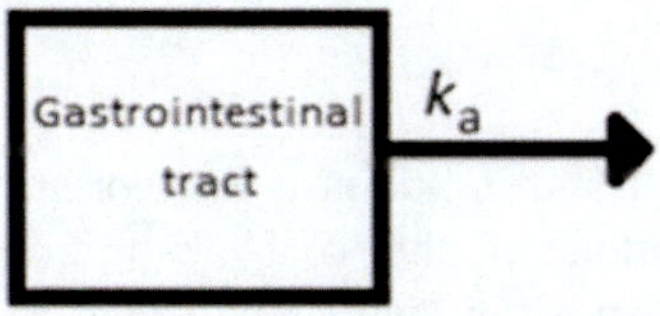

Fig. 2.1 A single absorption rate constant, k_a governs drug absorption throughout the gastrointestinal tract assuming a tank model. The sink conditions prevailing across the gastrointestinal membrane do not justify the mono-exponential decrease of drug quantity in the lumen A_{GI}, Eq. (2.5)

$$\frac{dC}{dt} = k_a \frac{FD}{V_d} e^{-k_a t} - k_{el} C \tag{2.6}$$

which upon integration eventually leads to the Bateman equation, Eq. (2.7):

$$C(t) = \frac{FDk_a}{V_d(k_a - k_{el})} \left(e^{-k_{el} t} - e^{-k_a t}\right) \tag{2.7}$$

Thus, *the infinite absorption time implied from* Eqs. (2.5) and (2.7) *results from the first-order change* (Eq. 2.4) *of the amount of drug in the gut lumen,* A_{GI}.

Since 1953, Eq. (2.7) is being used for the analysis of *C, t* data and the absorption rate constant k_a is routinely used as the parameter controlling the rate of drug absorption. Figure 2.1 shows the classical representation of the GI tract as a tank model. Given the anatomical-physiological complexity of the GI tract and the numerous drug processes, which are taking place in the lumen, the use of a single absorption rate constant oversimplifies the drug absorption phenomena.

However, the absorption rate constant k_a became the pharmacokinetic nemesis of all single and population PK–PD studies dealing with oral, pulmonary, and intranasal drug absorption.

In the next two sections, we are reviewing the biopharmaceutical and pharmacokinetic aspects of oral drug absorption placing emphasis on the absorption rate constant k_a, which is the common denominator of all these studies.

2.2 State of the Art in Oral Drug Absorption

To understand the current state of the art in oral drug absorption and the progress made over the years, we need to describe the early interest in this area. Two parallel but interconnected evolution routes were followed; the first started in 1953 (Sect. 2.2.1) emphasizing Pharmacokinetics (PK) and the second in the mid-1980s emphasizing Biopharmaceutics (Sect. 2.2.2). These two routes are interrelated.

2.2.1 The PK Route

As mentioned in Chap. 1, the starting point is the work of Dost and the adoption of Eq. (2.7) for the description of the oral absorption of drugs, assuming a one-compartment model disposition with first-order absorption and elimination rate. Despite the discrepancy associated with the infinite time of drug absorption, the absorption rate constant k_a became the gold standard as the most proper metric describing oral drug absorption. The routine use of k_a in all PK experimental studies can be seen in the publication of the late John Wagner's book in 1971 [1] and its computer version second edition [2] in 1993 whereas k_a was used in all compartmental models with first-order input.

In 1972, Lewis Sheiner, the founder of population PK, made the first step toward the modeling of individual PK for computer-aided drug dosage [3]. A few years later, Sheiner and Beal [4] developed the NONMEM (Non-Linear Mixed Effect Modelling) software, which is the basic tool for the estimation of the population pharmacokinetic parameters. These articles [3, 4] signify the birth of pharmacometrics, which relies on mathematical models describing the PK and pharmacodynamics (PD) of drugs, as well as the variability of the corresponding model parameters [5–7]. These advances resulted in a real explosion of population pharmacokinetic modeling work since the models developed [8, 9] quantify the information included in experimental or clinical data (concentration of a drug in blood, biomarkers, etc.) and summarize it in a limited set of parameters. Most importantly, statistical models are used to describe the variability of these parameters in populations of animals, volunteers, or patients. Again, the pharmacometric studies rely on the representation of oral absorption as a classical first-order process using an absorption rate constant (k_a). A variation of the classical model of absorption is a transit compartment model with a series of transit compartments with the same first-order transfer rate between each compartment in the GI lumen [10]. Although this adds flexibility to the fitting exercise, by varying the number of compartments in the GI lumen, the first-order notion is maintained [11]. It is widely understood that the commonly utilized models of drug absorption in population pharmacokinetics, which are mostly first order, with and without lag time or with transit compartments, often estimate large variabilities associated to k_a, which are unrealistic. Figure 2.2 shows the plot of citations for "the absorption rate constant" as a function of time in PUBMED [12]. The increase after 2005 is most likely associated with the explosion of pharmacometric studies and the development of relevant software packages close to the turn of the century.

2.2.1.1 Questioning the Absorption Rate Constant, k_a

The first theoretically justified questioning of the validity of k_a as a single parameter describing the absorption of drug throughout the time course of drug in the gastro-intestinal tract was published in 1997 [13]. The absorption processes, namely, drug

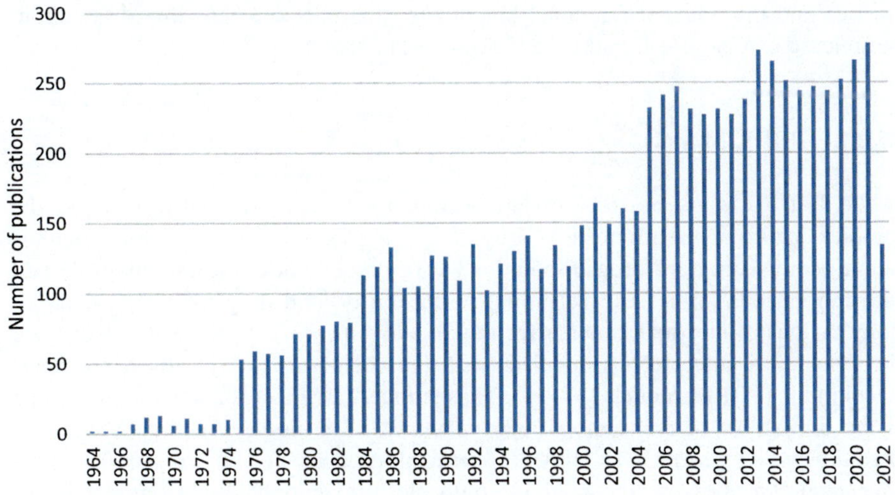

Fig. 2.2 Number of citations for "the absorption rate constant" per year in PUBMED [12] (accessed on 19/6/2022)

dissolution, transit, and uptake in the gastrointestinal tract were considered heterogeneous processes, since they take place at interfaces of different phases under variable stirring conditions. Given that the geometry of the environment is of major importance for the treatment of heterogeneous processes [13], fractal kinetics based on a time-dependent coefficient, k and not a rate constant was suggested as a better descriptor of the absorption kinetics [14],

$$k = k_1 t^{-h} \tag{2.8}$$

where h is different than zero fractal exponent of time t and k_1 is a constant expressed in $(\text{time})^{h\text{-}1}$ units. The value of h arises from two different phenomena: the heterogeneity (geometric disorder of the medium) and the imperfect mixing (diffusion limit) condition. Based on this analysis, drugs were classified in accordance with their gastrointestinal absorption characteristics into two broad categories, i.e., homogeneous and heterogeneous. The former category includes drugs with satisfactory solubility and permeability which ensure the validity of the homogeneous hypothesis (well-stirred homogenous media). Drugs with low solubility and permeability are termed heterogeneous, since they traverse the entire gastrointestinal tract and therefore are more likely to exhibit heterogeneous dissolution, transit, and uptake [14]. In the same vein, time-dependent absorption models were developed [15]. Besides, since the turn of the century, GastroPlus® is using the term "absorption rate coefficient" to describe the parameter controlling the drug absorption processes. Recently, instantaneous rate coefficients instead of rate constants have been proposed in modeling oral drug absorption [16]. Numerous applications of fractal kinetics and the relevant fractional kinetics to pharmacokinetics, drug dissolution, and release can be also found in the literature [17–36].

2.2.1.2 PBPK Models

In the last 20 years or so, biopharmaceutical scientists gradually unveiled the complex nature of gastrointestinal drug absorption phenomena, e.g., drug solubilization, supersaturation, drug dissolution, drug precipitation, the interplay of food with the various processes, (selective) permeability, and drug ionization changes along the gastrointestinal lumen affecting all above processes. In our days, all these phenomena are interpreted (modeled) with the use of physiologically based pharmacokinetic (PBPK) models [37–39]. These models are designed to overcome some of the limitations of *conventional compartmental PK models by integrating both drug-dependent parameters* (physicochemical properties) and *drug-independent parameters* (species-specific anatomy, physiology). Undoubtedly, a single value of the absorption rate constant cannot capture the complex dynamics of the absorption phase phenomena taking place concurrently. It is worthy to mention that drug absorption is assessed in PBPK studies using the permeability estimate expressed in velocity units (length/time). For the PBPK models, the duration of the absorption process, τ is a pivotal element. Thus, the user/modeler of the software packages (GastroPlus® Software, n.d.; Simcyp® Simulator, n.d.; PK-Sim® Software, n.d.) of the PBPK models fixes a finite time absorption period, e.g., 199 min [40], or transit times for each anatomical segment are specified [37–39]. In some cases [39], when PBPK models are coupled with a pharmacokinetic model, the fraction of dose absorbed is related proportionally to drug concentration in the gastrointestinal lumen since “the blood on the basolateral side of the membrane is regarded as an ideal sink.” Overall, the fixed time duration of the absorption processes and the deviations from the classical first-order absorption have been adopted in the PBPK models. Nevertheless, the PBPK models incorporate additional first-order processes to simulate gastrointestinal transit resulting in % absorption–time curves of typical first-order nature.

2.2.2 *The Bioph Route*

The important scientific and regukatory steps, which helped in the understanding of biopharmaceutical phenomena since the conception of the terms biopharmaceutics and pharmacokinetics in the 1950s and 1960s have been reviewed [41]. The absorption processes, drug dissolution, and drug uptake have been the subject of extensive research all these years [26, 37, 42]. The modeling work in oral drug absorption toward the estimation-prediction of a fraction of dose absorbed started in mid-1980s.

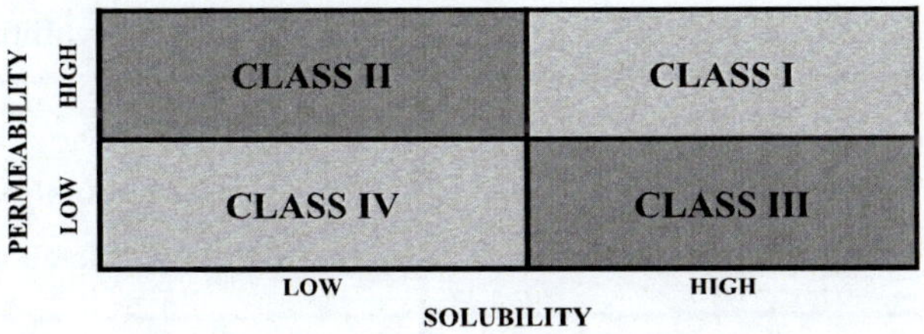

Fig. 2.3 The biopharmaceutic classification system (BCS). This modified version follows a Cartesian type of plot whereas a Class I (high solubility, high permeability) lies in the upper right corner as shown in [48]

The very first publications in the field were based on the absorption potential concept [43] and its quantitative version [44]. However, Professor Gordon Amidon's modeling work based on macroscopic [45] and microscopic [46] approaches using a tube model for the analysis of oral drug absorption revealed that two fundamental drug properties, namely, drug's solubility and permeability of GI membrane, determine the extent of oral drug absorption. This was followed by the formulation of the biopharmaceutic classification system (BCS) in 1995 whereas drugs are classified into four classes, i.e., I, II, III, and IV [47], Fig. 2.3.

The most important implication of these scientific advances is the publication of the two bioequivalence guidelines issued by the US Food & Drug Administration (FDA) [49] in 2000 (revised in 2017) and by European Medicines Agency [50] published in 2010. According to the guidelines a highly soluble, highly permeable drug (Class I) can get a biowaiver status for bioequivalence studies. This does not apply to Class II (low solubility, high permeability), and Class IV (low solubility, low permeability) drugs. For Class III (high solubility, low permeability) a biowaiver status can be assigned under certain conditions [49, 50]. Class I drugs exhibit extensive absorption (fraction of dose absorbed >0.90), while for Class II, III, and IV drugs the fraction of dose absorbed is certainly lower than 0.90.

In 2005, Wu and Benet [51] introduced a derivative classification system, the Biopharmaceutics Drug Disposition Classification System (BDDCS), after recognizing a strong association between the intestinal permeability and the extent of metabolism. Thus, drugs are classified into four BDDCS classes based on solubility and the extent of metabolism, Fig. 2.4.

Since in vitro and in vivo permeability estimates do not always predict the extent of oral drug absorption, BDDCS has gained regulatory acceptance [49, 50]. Figure 2.5 shows schematically in chronological order the major scientific and regulatory steps described above which contributed remarkably to our understanding of drug absorption phenomena. Although the development of PBPK models is not mentioned in Fig. 2.5, the first version of GastroPlus™ was introduced in 1998 and this can be considered as the starting point for the PBPK era.

	High Solubility	Low Solubility
Extensive Metabolism	**Class 1** High Solubility Extensive Metabolism	**Class 2** Low Solubility Extensive Metabolism
Poor Metabolism	**Class 3** High Solubility Poor Metabolism	**Class 4** Low Solubility Poor Metabolism

Fig. 2.4 The biopharmaceutic drug disposition classification system (BDDCS)

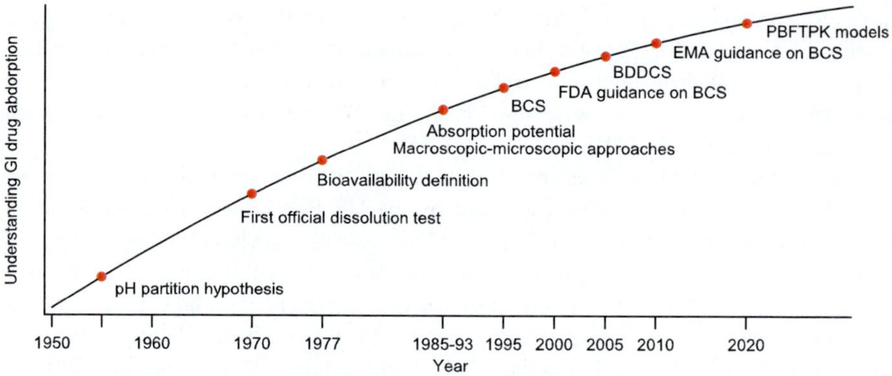

Fig. 2.5 A schematic of the important scientific and regulatory steps, which contributed to the understanding of biopharmaceutical phenomena. Key: 1950–1960: initial permeation solute studies based on pH-partition hypothesis (only the unionized nonpolar drug penetrates the membrane); 1970: *United States Pharmacopeia* (USP) published an official dissolution test in 12 monographs; 1977: FDA defines bioavailability as "the extent and rate to which the active drug ingredient or active moiety from the drug product is absorbed and becomes available at the site of drug action" on January 7, 1977; 1985–1993: the first approaches [43–46] for the estimation of fraction of dose absorbed were published; 1995: publication of BCS paper [47]; 2000: publication of FDA guidance on BCS [49]; 2005: publication of BDDCS paper [51]; 2010: publication of EMA guidance on BCS [50]; 2020: the Physiologically Based Finite Time Pharmacokinetic (PBFTPK) models were introduced (see Chaps. 3 and 4)

2.2.3 *The Absorption Rate Constant ($\mathbf{k}_a$) Links the Bioph and PK Routes*

An important result of the macroscopic [45] and microscopic [46] analyses of the gastrointestinal absorption using the tube model is Eq. (2.9), which relates the absorption rate constant (k_a) with the effective permeability (P_{eff}),

$$k_a = \frac{2P_{\mathrm{eff}}}{R} \tag{2.9}$$

where R refers to the radius of the small intestine. The last equation reveals that the parameters absorption rate constant and effective permeability are related proportionally; theoretically, one can get an estimate for k_a using Eq. (2.9) by measuring P_{eff} from the rate at which the drug permeates through a membrane, e.g., Caco II cells, in accord with Fick's first law of diffusion. The effective permeability (P_{eff}) estimate (expressed in velocity units, length/time) is used extensively as a metric of drug input rate in the PBPK models and its pH dependency is taken into account. However, the development of BDDCS [51] was based on the fact that the in vitro and in vivo permeability estimates do not always predict the extent of oral drug absorption; thus, the extent of metabolism is used instead of permeability in BDDCS. In parallel, the absorption rate constant has been found to be the most variable parameter in all pharmacokinetic studies. The take-home message is that neither k_a nor P_{eff} can be used as sole predictors of drug absorption due to the complex anatomic, physiologic, and hydrodynamic conditions prevailing in the gastrointestinal lumen.

2.3 The Unphysical Hypothesis of Infinite Absorption Time

Prelude In February 2019, one of the authors of the book (Panos Macheras) got up one morning and said to his wife "I think that we have been making a mistake since 1953." The result of this contemplation was the questioning of the unphysical hypothesis of the Bateman equation, i.e., drug absorption is taking place in infinite time [52]. Panos was fortunate to meet Dr. Athanasios (Thanos) Tsekouras shortly afterward. They first met in the early 1980s when Athanasios was a second year undergraduate Chemistry student specializing in interfacing analytical instruments with computers. At that time, Athanasios helped Panos extensively with computer-based pharmacokinetic simulations and teaching. A tiny piece of appreciation was published in the Acknowledgments section of the article [53] "The author wishes to thank Th. Tsekouras for his assistance in the computing work." According to Panos, Athanasios is a collaborator in every sense of the word. He is helpful, patient, caring, and above all unique and rare in computational and theoretical work. Today, Panos says that thanks to Athanasios this book came to fruition.

On the unphysical hypothesis of infinite absorption time [52] Based on common wisdom one can say that oral drug absorption takes place in a certain period of time, e.g., 0.5, 2, 10 h. In complete contrast, infinite time is required for the decay of nuclei analyzed by Bateman (Chap. 1). The finite time of absorption is indirectly included in the today's regulatory guidelines [49, 50]. For example, the rapid ($<$30 min) or very rapid ($<$15 min) regulatory dissolution criteria for Class I drugs imply rapid and complete absorption (fraction of dose absorbed $>$0.90). Consequently, one should

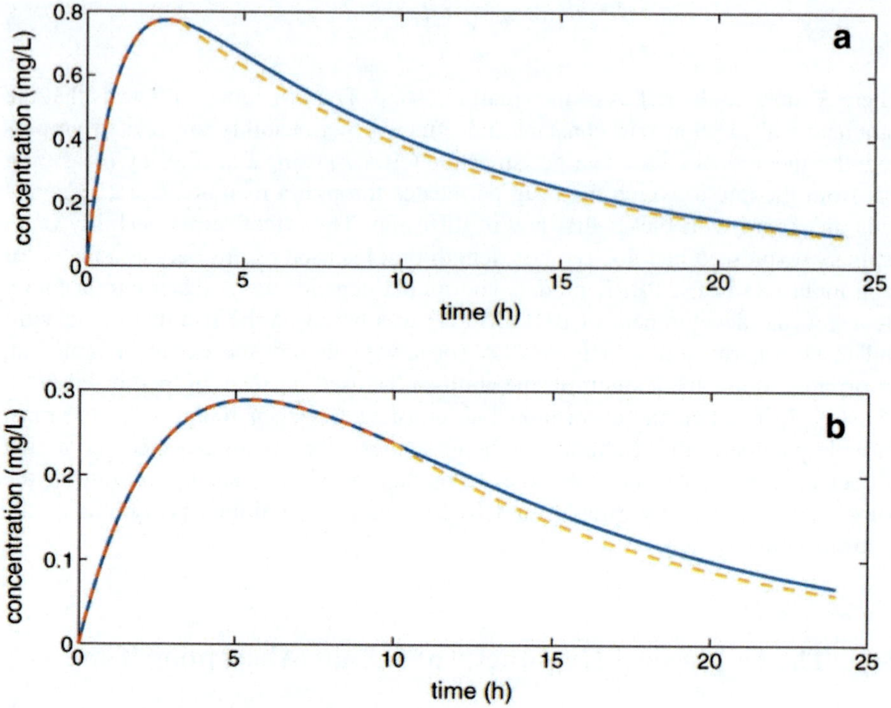

Fig. 2.6 Simulated curves generated from Eq. (2.7). Key: (**a**), continuous line, $F = 1$, $D = 100$ mg, $V = 100$ L, $k_a = 1$ h^{-1}, $k_{el} = 0.1$ h^{-1}; (**a**), dotted line, $F = 1$, $D = 100$ mg, $V = 100$ L, $k_a = 1$ h^{-1}, $k_{el} = 0.1$ h^{-1}, $\tau = 3$ h followed by mono-exponential disposition with $k_{el} = 0.1$ h^{-1}; (**b**), continuous line, $F = 0.5$, $D = 100$ mg, $V = 100$ L, $k_a = 0.3$ h^{-1}, $k_{el} = 0.1$ h^{-1}; (**b**), dashed line, $F = 0.5$, $D = 100$ mg, $V = 100$ L, $k_a = 0.3$ h^{-1}, $k_{el} = 0.1$ h^{-1}, $\tau = 10$ h followed by mono-exponential disposition with $k_{el} = 0.1$ h^{-1}

expect the termination of the absorption phase for a Class I drug at t_{max} (time which corresponds to the maximum concentration, C_{max}) because of complete drug absorption. On the contrary, the heterogeneous drugs [14] (Class II, III, and IV) which travel throughout the entire GI tract, exhibit usually slow and incomplete absorption (fraction of dose absorbed <0.90). Intuitively, the termination of the finite absorption phase for Class II, III, and IV drugs can be observed at longer times than Class I drugs, but certainly in all cases not beyond the *passage* of the unabsorbed (undissolved plus dissolved) amount of drug from the region of the absorptive sites of the GI.

Figure 2.6 shows simulated concentration-time curves for a rapidly (Fig. 2.6a) and a slowly (Fig. 2.6b) absorbed drug [52]. The curves have been generated from Eq. (2.7) (which is the Bateman equation, Eq. 1.2) assigning the finite absorption time, τ equal to 3 and 10 h for the rapidly and slowly absorbed drug, respectively. This is followed, by a mono-exponential elimination phase which starts at 3 and 10 h, respectively (Fig. 2.6, dotted lines). The corresponding curves generated from the Bateman equation (Eq. 2.7) using the same parameter values without the discontinuation of absorption phase are co-plotted in Fig. 2.6 (continuous curves). Visual inspection of Fig. 2.6 reveals that the descending limbs of the curves (continuous versus dotted lines) in a pair-wise comparison are quite similar. Obviously, when experimental data points are available and analyzed the discontinuation of the absorption phase followed by the elimination phase cannot be justified. Due to the curves' similarity, one can say that "this difference is not the end of the world." However, *there is a very important-crucial implication if the finite absorption time, τ is a valid hypothesis, namely, "*$[\mathrm{AUC}]_0^{\tau}$ *and not* $[\mathrm{AUC}]_0^{\infty}$ *is the proper metric for the extent of drug absorption.*" If this is so, the physiologically sound finite absorption time has been misinterpreted since the beginning of pharmacokinetics in 1953. Relying on the above syllogisms and the simple example of Fig. 2.6, one can conclude that the wrong adoption by Dost [54] of Eq. (2.7) for the analysis of oral data seriously affected fundamental aspects of biopharmaceutics and pharmacokinetics.

In this vein, the integral of Eq. (2.7) from zero to infinity, which corresponds to area under the curve, $[\mathrm{AUC}]_0^{\infty}$, has been established as a measure of drug *extent of absorption; it is routinely used in pharmacokinetics* as a metric of drug's extent of absorption because of the proportionality $[\mathrm{AUC}]_0^{\infty} = \mathrm{FD}/(k_{\mathrm{el}}V)$ (see Chap. 1, Eq. 1.4). Obviously, this simple relationship is inherently linked with the infinite time for the duration of both first-order processes, namely, drug absorption and elimination. However, the present considerations indicate that the $[\mathrm{AUC}]_0^{\tau}$ (where τ denotes the finite duration of the absorption phase) is the proper indicator for the bioavailable fraction of drug. Mathematically, $[\mathrm{AUC}]_0^{\tau}$ can be calculated [55] from Eq. (2.10) assuming one-compartment model with first-order absorption and elimination:

$$\int_0^{\tau} C dt = [\mathrm{AUC}]_0^{\tau} = \frac{\mathrm{FD}}{V k_{\mathrm{el}}}\left(1 - \frac{k_a e^{-k_{\mathrm{el}}\tau} - k_{\mathrm{el}} e^{-k_a \tau}}{k_a - k_{\mathrm{el}}}\right) \tag{2.10}$$

Equation (2.10) reveals that the proportionality between $[\mathrm{AUC}]_0^{\tau}$ and the bioavailable fraction (FD) is maintained while both rate constants, k_{a}, k_{el} are important too. However, $[\mathrm{AUC}]_0^{\tau}$ *is an ideal metric for the extent of absorption by its own merit since it corresponds to the end of drug absorption at time* τ.

The analysis of C, t data based on the concept of the finite absorption time, requires the development of software for the estimation of τ. The corresponding datum point $(C(\tau), \tau)$ is a discontinuity point of a piecewise function defined by two sub-functions, i.e., for one-compartment model drugs Eq. (2.7) holds for $t \leq \tau$ and the mono-exponential elimination holds for $t > \tau$. In this context, one has to explore through simulation the issue as to under what circumstances, e.g., experimental error, $k_{\mathrm{a}}/k_{\mathrm{el}}$ ratios, it would be possible to observe or identify τ from the experimental data. Besides, other concepts associated with Eq. (2.7) require reconsideration. In this vein, the concept of flip-flop kinetics (Eq. 2.7 with $k_{\mathrm{a}} < k_{\mathrm{el}}$) and the equality of rate constants (see Eq. 1.3) become questionable since drug absorption for infinite time is associated with Eqs. (2.7) and (1.3). Similarly, other aspects of GI absorption analysis, e.g., the percent amount absorbed as a function of time plots, frequently used in in vitro–in vivo correlations, require different treatment under the premise of the finite absorption time. For example, the estimation of τ in oral drug absorption studies using various species can help or even improve the well-known poor interspecies bioavailability correlations. One of the reasons can be the remarkable difference in GI transit time. Hopefully, this work will initiate studies toward this end.

2.4 Concluding Remarks

It can be concluded that Einstein's quote in the subtitle of this chapter is very appropriate. The establishment of a unique first-order rate constant controlling oral drug absorption throughout the drug transit in the lumen, oversimplifies the complex drug absorption phenomena. This observation brings us to another insightful quote from Albert Einstein: "The important thing is not to stop questioning. Curiosity has its own reason for existing." The next chapter is based on the curiosity to unveil the mysteries of drugs' finite absorption time.

References

1. Wagner JG (1971) Biopharmaceutics and relevant pharmacokinetics. Drug Intelligence Publications, Hamilton
2. Wagner JG (1993) Pharmacokinetics for the pharmaceutical scientist. Technomic Publishing, Lancaster
3. Sheiner LB, Rosenberg B, Melmon KL (1972) Modelling of individual pharmacokinetics for computer-aided drug dosage. Comput Biomed Res 5:441–459. https://doi.org/10.1016/0010-4809(72)90051-1
4. Sheiner LB, Beal SL (1979) NONMEM users guide. Division of Pharmacology: University of California, San Francisco
5. Sheiner LB, Beal SL (1980) Evaluation of methods for estimating population pharmacokinetics parameters. I. Michaelis-Menten model: routine clinical pharmacokinetic data. J Pharmacokin Biopharm 8:553–571. https://doi.org/10.1007/BF01060053
6. Sheiner LB, Beal SL (1981) Evaluation of methods for estimating population pharmacokinetic parameters. II. Biexponential model and experimental pharmacokinetic data. J Pharmacokin Biopharm 9:635–651. https://doi.org/10.1007/BF01061030
7. Sheiner LB, Beal SL (1983) Evaluation of methods for estimating population pharmacokinetic parameters. III. Monoexponential model: routine clinical pharmacokinetic data. J Pharmacokin Biopharm 11:303–319. https://doi.org/10.1007/BF01061870
8. Bonate P (2011) Pharmacokinetic-pharmacodynamic modeling and simulation, 2nd edn. Springer US, Boston
9. Mould DR, Upton RN (2013) Basic concepts in population modeling, simulation, and model-based drug development—part 2: introduction to pharmacokinetic modeling methods. CPT: Pharmacometrics Syst Pharmacol 2:e38. https://doi.org/10.1038/psp.2013.14

10. Savic RM, Jonker DM, Kerbusch T, Karlsson MO (2007) Implementation of a transit compartment model for describing drug absorption in pharmacokinetic studies. J Pharmacokinet Pharmacodyn 34:711–726. https://doi.org/10.1007/s10928-007-9066-0
11. Hénin E, Bergstrand M, Standing JF, Karlsson MO (2012) A mechanism-based approach for absorption modeling: the gastro-intestinal transit time (GITT) model. AAPS J 14:155–163. https://doi.org/10.1208/s12248-012-9324-y
12. US National Library of Medicine. https://pubmed.ncbi.nlm.nih.gov/
13. Kopelman R (1988) Fractal reaction kinetics. Science 241:1620–1626. https://doi.org/10.1126/science.241.4873.1620
14. Macheras P, Argyrakis P (1997) Gastrointestinal drug absorption: is it time to consider heterogeneity as well as homogeneity? Pharm Res 14:842–847. https://doi.org/10.1023/A:1012183313218
15. Higaki K, Yamashita S, Amidon GL (2001) Time-dependent oral absorption models. J Pharmacokinet Pharmacodyn 28:109–128. https://doi.org/10.1023/a:1011573831444
16. Mavroudis P, Kosmidis K, Macheras P (2019) On the unphysical hypotheses in pharmacokinetics and oral drug absorption: time to utilize instantaneous rate coefficients instead of rate constants. Eur J Pharm Sci 130:137–146. https://doi.org/10.1016/j.ejps.2019.01.027
17. Macheras P (1996) A fractal approach to heterogeneous drug distribution: calcium pharmacokinetics. Pharm Res 13:663–670. https://doi.org/10.1023/A:1016031129053
18. Macheras P, Dokoumetzidis A (2000) On the heterogeneity of drug dissolution and release. Pharm Res 17:108–112. https://doi.org/10.1023/A:1007596709657
19. Rinaki E, Valsami G, Macheras P (2003) The power law can describe the "entire" release curve: a hypothesis. Int J Pharm 255:199–207. https://doi.org/10.1016/S0378-5173(03)00079-6
20. Kosmidis K, Argyrakis P, Macheras P (2003) A re-appraisal of drug release laws using Monte-Carlo simulations: the prevalence of the Weibull function. Pharm Res 20:988–995. https://doi.org/10.1023/A:1024497920145
21. Kosmidis K, Argyrakis P, Macheras P (2003) Fractal kinetics in drug release from finite fractal matrices. J Chem Phys 119:6373–6377. https://doi.org/10.1063/1.1603731
22. Dokoumetzidis A, Karalis V, Iliadis A, Macheras P (2004) The heterogeneous course of drug transit through the body. Trends Pharmacol Sci 25:140–146. https://doi.org/10.1016/j.tips.2004.01.008
23. Kosmidis K, Karalis V, Argyrakis P, Macheras P (2004) Michaelis-Menten kinetics under spatially constrained conditions: application to mibefradil pharmacokinetics. Biophys J 87:1498–1506. https://doi.org/10.1529/biophysj.104.042143
24. Papadopoulou V, Kosmidis K, Vlachou M, Macheras P (2006) On the use of the Weibull function for the discernment of drug release mechanisms. Int J Pharm 309:44–50. https://doi.org/10.1016/j.ijpharm.2005.10.044
25. Dokoumetzidis A, Papadopoulou V, Macheras P (2006) Analysis of dissolution data using modified versions of Noyes-Whitney equation and the Weibull function. Pharm Res 23:256–261. https://doi.org/10.1007/s11095-006-9093-3
26. Dokoumetzidis A, Macheras P (2006) A century of dissolution research: from Noyes and Whitney to the biopharmaceutics classification system. Int J Pharm 321:1–11. https://doi.org/10.1016/j.ijpharm.2006.07.011
27. Pang KS, Weiss M, Macheras P (2007) Advanced pharmacokinetic models based on drug clearance, circulatory and fractal concepts. Am Ass Pharm Sci J 9:E268–E283. https://doi.org/10.1208/aapsj0902030
28. Kosmidis K, Macheras P (2007) Monte Carlo simulations for the study of drug release from matrices with high and low diffusivity areas. Int J Pharm 343:166–172. https://doi.org/10.1016/j.ijpharm.2007.05.021
29. Kosmidis K, Macheras P (2008) Monte Carlo simulations of drug release from matrices with periodic layers of high and low diffusivity. Int J Pharm 354:111–116. https://doi.org/10.1016/j.ijpharm.2007.10.036

30. Dokoumetzidis A, Macheras P (2009) Fractional kinetics in drug absorption and disposition processes. J Pharmacokinet Pharmacodyn 36:165–178. https://doi.org/10.1007/s10928-009-9116-x
31. Kytariolos J, Dokoumetzidis A, Macheras P (2010) Power law IVIVC: an application of fractional kinetics for drug release and absorption. Eur J Pharm Sci 41:299–304. https://doi.org/10.1016/j.ejps.2010.06.015
32. Dokoumetzidis A, Magin R, Macheras P (2010) Fractional kinetics in multi-compartmental systems. J Pharmacokinet Pharmacodyn 37:1–18
33. Dokoumetzidis A, Kosmidis K, Macheras P (2011) Monte Carlo simulations and fractional kinetics considerations for the Higuchi equation. Int J Pharm 418:100–103. https://doi.org/10.1016/j.ijpharm.2010.11.046
34. Dokoumetzidis A, Macheras P (2011) The changing face of the rate concept in biopharmaceutical sciences: from classical to fractal and finally to fractional. Pharm Res 28:1129–1132. https://doi.org/10.1007/s11095-011-0370-4
35. Sopasakis P, Sarimveis C, Macheras P, Dokoumetzidis A (2018) Fractional calculus in pharmacokinetics. J Pharmacokinet Pharmacodyn 45:107–125. https://doi.org/10.1007/s10928-017-9547-8
36. Kosmidis K, Macheras P (2018) On the dilemma of fractal or fractional kinetics in drug release studies: a comparison between Weibull and Mittag-Leffler functions. Int J Pharm 543:269–273. https://doi.org/10.1016/j.ijpharm.2018.03.060
37. Charalabidis A, Sfouni M, Bergström C, Macheras P (2019) The biopharmaceutics classification system (BCS) and the biopharmaceutics drug disposition classification system (BDDCS): beyond guidelines. Int J Pharm 566:264–281. https://doi.org/10.1016/j.ijpharm.2019.05.041
38. Krzyzanski W, Jusko WJ, Wacholtz MC, Minton N, Cheung WK (2005) Pharmacokinetic and pharmacodynamic modeling of recombinant human erythropoietin after multiple subcutaneous doses in healthy subjects. Eur J Pharm Sci 26:295–306. https://doi.org/10.1016/j.ejps.2005.06.010
39. Sjögren E, Westergren J, Grant I, Hanisch G, Lindfors L, Lennernäs H, Abrahamsson B, Tannergren C (2013) In silico predictions of gastrointestinal drug absorption in pharmaceutical product development: application of the mechanistic absorption model GI-Sim. Eur J Pharm Sci 49:679–698. https://doi.org/10.1016/j.ejps.2013.05.019
40. Macheras P, Karalis V (2014) A non-binary biopharmaceutical classification of drugs: the ΑΒΓ system. Int J Pharm 464:85–90. https://doi.org/10.1016/j.ijpharm.2014.01.022
41. Macheras P, Karalis V, Valsami G (2013) Keeping a critical eye on the science and the regulation of oral drug absorption: a review. J Pharm Sci 102:3018–3036. https://doi.org/10.1002/jps.23534
42. Sugano K (2012) Biopharmaceutics modeling and simulations: theory, practice, methods, and applications. Wiley, Hoboken
43. Dressman JB, Amidon GL, Fleisher D (1985) Absorption potential: estimating the fraction absorbed for orally administered compounds. J Pharm Sci 74:588–589. https://doi.org/10.1002/jps.2600740523
44. Macheras P, Symillides M (1989) Toward a quantitative approach for the prediction of the fraction of dose absorbed using the absorption potential concept. Biopharm Drug Disp 10:43–53. https://doi.org/10.1002/bdd.2510100106
45. Sinko PJ, Leesman GD, Amidon GL (1991) Predicting fraction dose absorbed in humans using a macroscopic mass balance approach. Pharm Res 8:979–988. https://doi.org/10.1023/A:1015892621261
46. Oh D-M, Curl RL, Amidon GL (1993) Estimating the fraction dose absorbed from suspensions of poorly soluble compounds in humans: a mathematical model. Pharm Res 10:264–270. https://doi.org/10.1023/A:1018947113238
47. Amidon GL, Lennernäs H, Shah VP, Crison JR (1995) A theoretical basis for a biopharmaceutic drug classification: the correlation of in vitro drug product dissolution and in vivo bioavailability. Pharm Res 12:413–420. https://doi.org/10.1023/a:1016212804288

48. Macheras P, Iliadis A (2016) Oral drug absorption. In: Modeling in biopharmaceutics, pharmacokinetics and pharmacodynamics: homogeneous and heterogeneous approaches, 2nd edn. Springer, Cham, pp 483
49. Food and Drug Administration (2017) Waiver of in vivo immediate-release solid oral bioequivalence studies for bioavailability and biopharmaceutics classification dosage forms based on biopharmaceutics classification dosage forms based on a Biopharmaceutics Classification System Guidance For. https://www.fda.gov/media/70963/download
50. EMA. Guideline on the investigation of bioequivalence. Published online December 2010: 27. doi:Doc. Ref.: CPMP/EWP/QWP/1401/98 Rev. 1/ Corr ** COMMITTEE
51. Wu C-Y, Benet LZ (2005) Predicting drug disposition via application of BCS: transport/absorption/elimination interplay and development of a biopharmaceutics drug disposition classification system. Pharm Res 22:11–23. https://doi.org/10.1007/s11095-004-9004-4
52. Macheras P (2019) On an unphysical hypothesis of Bateman equation and its implications for pharmacokinetics. Pharm Res 36:94. https://doi.org/10.1007/s11095-019-2633-4
53. Macheras P (1987) Method of residuals: estimation of absorption and elimination rate constants having comparable values. Biopharm Drug Disp 8:47–56. https://doi.org/10.1002/bdd.2510080106
54. Dost FH (1953) Der Blutspiegel. Kinetik der Konzentrationsabläufe in der Kreislaufflüssigkeit. Thieme, Leipzig
55. Niazi S (1979) Textbook of biopharmaceutics and clinical pharmacokinetics. Appleton-Century-Crofts, New York

Chapter 3
The Finite Absorption Time (FAT) Concept

Nothing pertaining to humanity becomes us so well as mathematics.
Isaac Asimov (1920–1992)

Contents

Abstract A minimal model of gastrointestinal drug absorption is described. The assumptions of the model are as follows: (1) Drugs are absorbed passively under sink conditions. (2) The limiting property is solubility and/or permeability in accordance with the drug's BCS classification. (3) Physiological time limits were applied for drug absorption in the small intestines or colon. Differential equations for drug's permeation rate were written taking into account the drug's BCS classification. Simulations were carried out using the model equations. Fittings of these equations were also performed on experimental data.

Keywords Finite absorption time · BCS · Cephradine · Ibuprofen · Flurbiprofen · Itraconazole

P. Macheras, A. A. Tsekouras, *Revising Oral Pharmacokinetics, Bioavailability and Bioequivalence Based on the Finite Absorption Time Concept*,
https://doi.org/10.1007/978-3-031-20025-0_3

3.1 Sink Conditions Imply Zero-Order Kinetics

In this section, we establish the need for and prove the validity of zero-order kinetics in oral drug absorption. The basic concept is that drugs are absorbed under sink conditions implying zero-order kinetics. In this context, the first ever zero-order problem was the one proposed by Metrodorus among many simple mathematical problems.

Panos Macheras was introduced to this puzzle by his graduate advisor, Dr. Arnold Rosen, a professor at Chelsea College (now King's College), University of London, known for his work on estimation of drug–protein binding constants. He explained zero-order kinetics using a Metrodorus riddle, perhaps, the first in history hydraulic compartmental model relevant to the elimination processes. Zero-order kinetics can be encountered in carrier-mediated transport or Michaelis–Menten elimination kinetics under saturable conditions.

We replicate this riddle here for its educational merits. The problem as stated by Metrodorus goes as follows:

> Κύκλωψ † ἐγὼ Πολύφημος ὁ χάλκεος: οἷα δ᾽ ἐπ᾽ αὐτῷ
> τεῦξέ τις ὀφθαλμὸν καὶ στόμα καὶ παλάμην,
> κρουνοῖς συζεύξας: στάζοντι δὲ πάμπαν ἔοικεν,
> ἠδ᾽ ἔτι καὶ βλύζων φαίνετ᾽ ἀπὸ στόματος.
> κρουνῶν δ᾽ οὔτις ἄτακτος: ὁ μὲν παλάμης τρισὶ μούνοις
> ἤμασιν ἐμπλήσει δεξαμενὴν προρέων
> ἠμάτιος γλήνης: στόμα δ᾽ ἤματος ἐν δύο πέμπτοις.
> τίς κ᾽ ἐνέποι τρισσοῖς ἶσα θέοντα χρόνον;

This is Polyphemus the brazen Cyclops, and as if on him someone made an eye, a mouth, and a hand, connecting them with pipes. He looks quite as if he were dripping water and seems also to be spouting it from his mouth. None of the spouts are irregular; that from his hand when running will fill the cistern in 3 days only, that from his eye in 1 day, and his mouth in two-fifths of a day. Who will tell me the time it takes when all three are running?

[Source of translation: 1918: W.R. Paton: The Greek Anthology Book XIV. Metrodorus: 132, Arithmetical Epigram of Metrodorus https://proofwiki.org/wiki/Greek_Anthology_Book_XIV:_Metrodorus:_132].

Setting up the problem and solving the riddle is straightforward. According to the text

$T_1 = 1$ day, $T_2 = \frac{2}{5}$ day, $T_3 = 3$ days, hence $k_1 = \frac{D}{T_1}$, $k_2 = \frac{D}{T_2}$, $k_3 = \frac{D}{T_3}$, where k_1, k_2, and k_3 are the rate constants for each water outlet and D is the capacity of the cistern. If $A(t)$ is the cistern contents at time t, then the differential equation describing the problem is the following:

$$\frac{\mathrm{d}A(t)}{\mathrm{d}t} = k_1 + k_2 + k_3$$

the summation indicating the additivity of zero-order processes, Fig. 3.1.

Upon integration we get:

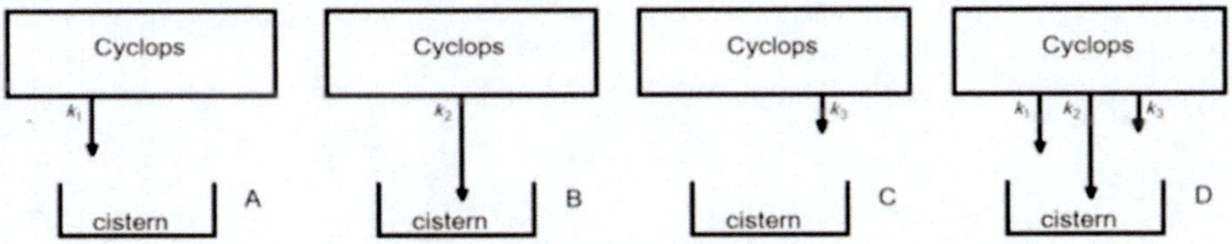

Fig. 3.1 Schematic of the Metrodorus riddle. Cases A, B, and C show single inputs to the cistern, while case D shows concurrent delivery from all three inputs

$$\int_0^t \frac{\mathrm{d}A(t)}{\mathrm{d}t}\mathrm{d}t = \int_0^t (k_1 + k_2 + k_3)\mathrm{d}t$$

$$A(t) - A(0) = (k_1 + k_2 + k_3)t$$

If we start with an empty cistern and at time t we have a full cistern, then $A(0) = 0$ and $A(t) = D$, hence

$$t = \frac{D}{k_1 + k_2 + k_3} = \frac{1}{\frac{1}{T_1} + \frac{1}{T_2} + \frac{1}{T_3}} = \frac{1}{\frac{1}{1} + \frac{5}{2} + \frac{1}{3}}\text{days} = \frac{6}{23}\text{days}$$

This simple example demonstrates the "additivity" characteristic of the zero-order processes, namely, when the causative factors act together like the three flow rates the producing effect is the sum of their individual effects. The same principle is applied in gastrointestinal drug absorption. It follows zero-order kinetics as a result of the sink conditions prevailing due to the high blood flow in the vena cava; however, elimination and distribution processes of first-order character or Michaelis–Menten kinetics are concurrently taking place with the zero-order process(es) of drug absorption.

3.2 Finite Absorption Time

Upon the publication [1] of the commentary on the unphysical assumption of infinite absorption time, a minimal drug absorption model was developed [2] relying on the Finite Absorption Time (FAT) concept. Figure 3.2 shows the basic characteristics of the model in terms of the biopharmaceutical drug properties and the physiological transit times in the gastrointestinal tract [3]. The high blood flow in the vena cava [4] imposes sink conditions in the neighborhood of the gastrointestinal membrane, which implies zero-order absorption kinetics.

According to Fig. 3.2, drug absorption takes place in the physiological limits for the intestinal and colon transit times of 5 and 30 h, respectively [3]. Intuitively, absorption for Class I drugs is completed in the intestines, while Class II, III, and IV drugs' absorption can be also observed in the colon to a lesser degree, Fig. 3.2. In

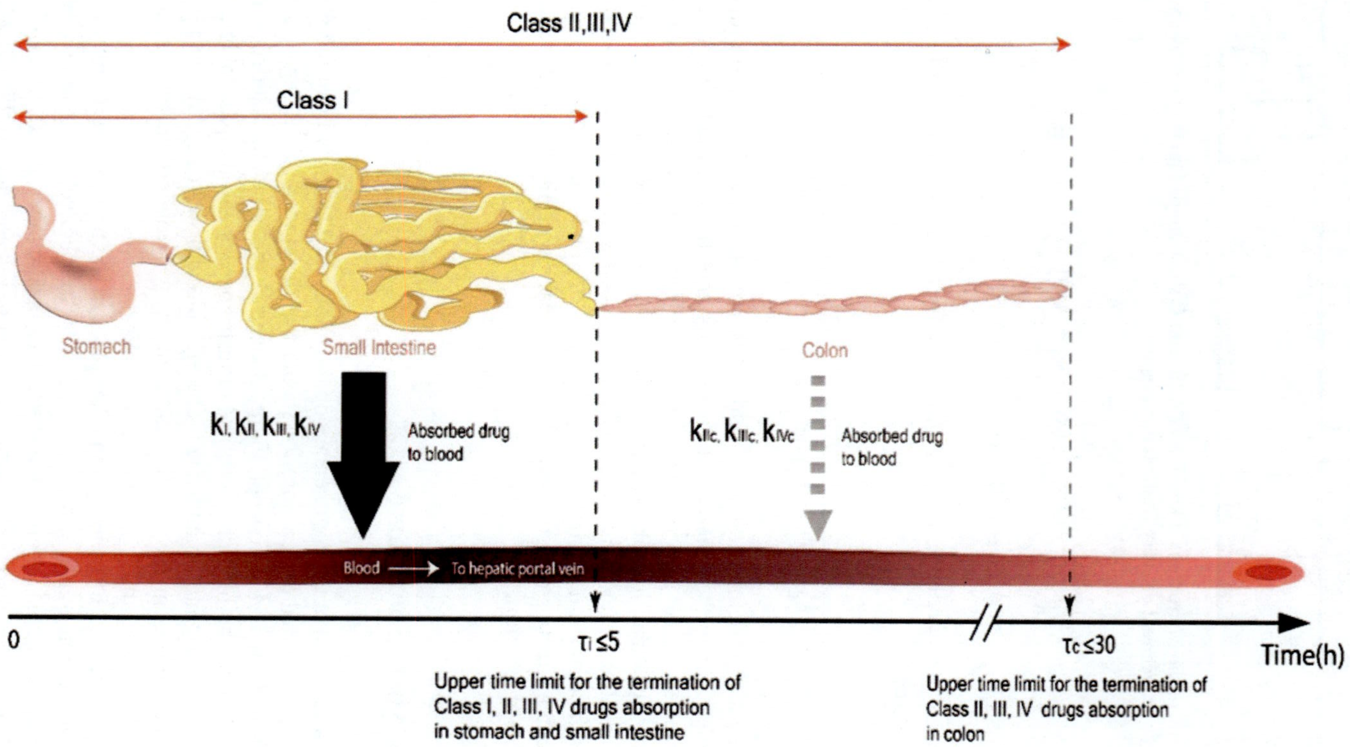

Fig. 3.2 A schematic of the biopharmaceutical/physiological drug absorption model, which relies on the transit times of the drug along the gastrointestinal tract. For Class I drugs, the completion of absorption ($F > 0.90$) ceases in a shorter time than the duration of the stomach and small intestine transit 4.86 h [3]. For Class II, III, and IV drugs, the limited overall absorption ($F < 0.90$) can be continued beyond the ileocecal valve and lasts not more than the whole gut transit time, e.g., 29.81 h [3]. The absorbed drug reaches the hepatic portal vein, while the blood flow (20–40 cm/s) [4] imposes sink conditions on drug transfer. The thick black arrow denotes the major site of drug absorption, namely, the small intestine. The dashed arrow indicates the potentially limited drug absorption from the colon

this vein, the FAT for Class I drugs should be much smaller than 5 h; the absorption of Class II, III, and IV drugs should be also terminated in less than 5 h since the small intestines are the region with the drug's absorptive sites; however, absorption for Class II, III, and IV drugs can also take place in the colon, but not beyond the physiological time limit of 30 h [3].

3.2.1 *Modeling Work: Combining BCS with Absorption Kinetics*

The rate of penetration for any drug obeying passive absorption under sink conditions can be expressed with Eq. (3.1).

$$\text{Rate of penetration} = P \cdot (\text{SA})(C_{\text{GI}}) \tag{3.1}$$

where P is the permeability of the drug expressed in velocity units (length/time), SA is the surface area of the gastrointestinal membrane in (length)^2 units, and C_{GI} is the drug concentration in the GI tract. Equation (3.1) can be applied to drugs with different biopharmaceutical properties.

Class I Drugs For highly soluble, highly permeable drugs (Class I), the rate of permeation is high. Regardless of the formulation administered (drug solution or solid formulation), these drugs do not exhibit either dissolution or permeability limited absorption. Therefore the rate of penetration of a Class I drug whose absorption is completed in the small intestines can be written as:

$$(\text{Rate of Penetration})_{\text{I}} = P \cdot (\text{SA})_{\text{i}} \cdot C_{\text{GI}} = k_{\text{I}} = \frac{F_{\text{i}} D}{\tau_{\text{i}}} = \frac{D}{\tau_{\text{i}}} \tag{3.2}$$

where k_{I} denotes the constant penetration rate (mass/time units) for Class I drugs, F_{i} is the fraction of dose absorbed in the stomach and small intestine and τ_{i} is the duration of the initial absorption phase. Since Class I drugs are absorbed fully, $F_{\text{i}} = 1$ is being used in Eq. (3.2). Accordingly, the change of drug blood concentration C as a function of time for Class I drugs assuming one-compartment model disposition is:

$$V \frac{\text{d}C}{\text{d}t} = k_{\text{I}} - k_{\text{el}} \text{CV}_{\text{d}} = \frac{D}{\tau_{\text{i}}} - k_{\text{el}} \text{CV}_{\text{d}} \tag{3.3}$$

Plausibly, the small intestine is the major site of absorption for Class I drugs, while absorption always ceases in a much shorter time than 4.86 h, which is the sum of gastric and small intestine transit time [3], Fig. 3.2. Equation (3.3) gives upon integration for $t = 0$, $C = 0$ and $t = t$, $C = C$:

$$C(t) = \frac{D}{\tau_i}\frac{1}{V_d k_{el}}\left(1 - e^{-k_{el}t}\right) \tag{3.4}$$

Upon completion of the absorption phase at time $t = \tau_i$, the drug concentration will be $C(\tau_i)$ in accordance with Eq. (3.4). The change of drug concentration beyond time τ_i is described by the following equation:

$$\frac{dC}{dt} = -k_{el}C \tag{3.5}$$

Equation (3.5) upon integration for $t = \tau_i$, $C{=}C(\tau_i)$ and $t \to \infty$, $C = 0$, leads to Eq. (3.6) which describes the monotonic elimination phase.

$$C(t) = C(\tau_i)e^{-k_{el}(t-\tau_i)} \tag{3.6}$$

Class II Drugs For low soluble, highly permeable drugs (Class II), the rate of drug permeation is low, Eq. (3.1). This is so, since the maximum value of the factor C_{GI} in Eq. (3.1) cannot be higher than the low saturation solubility, C_S, of the drug in the gastrointestinal fluids. This solubility value can be also considered constant. Therefore, the rate of gastric and small intestine penetration for a Class II drug can be approximated:

$$(\text{Rate of penetration})_{II} = P \cdot (\text{SA})_i (C_S) = k_{II} = \frac{F_i D}{\tau_i} \tag{3.7}$$

where k_{II} denotes the constant penetration rate (mass/time units) for Class II drugs, Fig. 3.2. Accordingly, the change of drug blood concentration C as a function of time assuming one-compartment model disposition for Class II drugs is:

$$V\frac{dC}{dt} = k_{II} - k_{el}CV_d \tag{3.8}$$

Equations (3.7) and (3.8) roughly operate for not more than 4.86 h, which is the sum of gastric and small intestine transit time [3]. The passage of Class II drugs to the colon via the ileocecal valve, which separates the small intestine and the large intestine, can either result in the termination of drug absorption or the significant reduction of the rate of drug penetration since the effective surface area $(\text{SA})_c$ is much smaller in the colon and the amount of unabsorbed drug at the ileocecal valve is equal to $(1\text{-}F_i)D$:

$$(\text{Rate of penetration})_{II,c} = P \cdot (\text{SA})_c (C_S) = k_{II,c} = \frac{(1 - F_i)D}{\tau_c - \tau_i} \tag{3.9}$$

where τ_c denotes the termination time of drug absorption from the colon, $(\text{SA})_c$ is the surface area of the colon and $k_{II,c}$ denotes the constant penetration rate (mass/time units) for Class II drugs in the colon, Fig. 3.2. Accordingly, the change of drug blood concentration C as a function of time assuming one-compartment model disposition for Class II drugs during the drug passage through the colon is:

$$V_{\mathrm{d}} \frac{\mathrm{d}C}{\mathrm{d}t} = k_{\mathrm{II,c}} - k_{\mathrm{el}} C V_{\mathrm{d}} \tag{3.10}$$

This equation roughly holds from 4.86 h to the time needed for the drug to reach the non-absorptive sites of the colon, τ_c, but certainly shorter than 20.28 or 31.95 h, i.e., the colon transit time for a single- or multi-unit formulation, respectively [3], Fig. 3.2. At time τ_c absorption ceases; beyond this time point the drug is only eliminated from the body. Hence, the drug concentration decreases according to Eq. (3.11), which is similar to Eq. (3.6):

$$C(t) = C(\tau_c) e^{-k_{\mathrm{el}}(t - \tau_c)} \tag{3.11}$$

where $C(\tau_c)$ is the drug concentration corresponding to time τ_c.

Class III Drugs For highly soluble, low permeable drugs (Class III), the rate of drug permeation is low, Eq. (3.1). This is so, since the low permeability value, P_l, is rate limiting for absorption; therefore, the rate of penetration for a Class III drug, throughout the passage of drug from the stomach and small intestine, can be approximated:

$$(\text{Rate of penetration})_{\mathrm{III}} = P_l \cdot (\mathrm{SA})_i (C_S) = k_{\mathrm{III}} = \frac{F_i D}{\tau_i} \tag{3.12}$$

where k_{III} denotes the constant penetration rate (mass/time units) for Class III drugs, Fig. 3.2. Accordingly, the change of drug blood concentration C as a function of time for Class III drugs is:

$$V_{\mathrm{d}} \frac{\mathrm{d}C}{\mathrm{d}t} = k_{\mathrm{III}} - k_{\mathrm{el}} C V_{\mathrm{d}} \tag{3.13}$$

Equations (3.12) and (3.13) roughly operate for not more than 4.86 h, which is the sum of gastric and small intestine transit time [3]. The passage of Class III drugs to the colon via the ileocecal valve can either result in the termination of drug absorption or the significant reduction of the rate of drug penetration since the effective surface area $(\mathrm{SA})_c$ is much smaller in the colon and the amount of unabsorbed drug at the ileocecal valve is equal to $(1 - F_i)D$.

$$(\text{Rate of penetration})_{\mathrm{III,c}} = P_l \cdot (\mathrm{SA})_c (C_{\mathrm{GI}}) = k_{\mathrm{III,c}} = \frac{(1 - F_i)D}{\tau_c - \tau_i} \tag{3.14}$$

where $k_{\mathrm{III,c}}$, denotes the zero-order penetration rate (mass/time units) for Class III drugs in the colon. Accordingly, the change of drug blood concentration C as a function of time assuming one-compartment model disposition for Class III drugs in the colon is:

$$V_d \frac{dC}{dt} = k_{III,c} - k_{el} C V_d \tag{3.15}$$

This equation roughly holds from 4.86 h to the time needed for the drug to reach the non-absorptive sites of the colon, τ_c, but certainly shorter than 20.28 or 31.95 h, i.e., the colon transit time for a single- or multi-unit formulation, respectively [3]. At time τ_c absorption ceases; beyond this time point, the drug is only eliminated from the body. Hence, the drug concentration decreases according to Eq. (3.11) for $t \geq \tau_c$.

Class IV Drugs For low soluble, low permeable (Class IV) drugs, the rate of permeation is low, Eq. (3.1). Both solubility and permeability are limiting absorption. The low values of the terms P and C_{GI} in Eq. (3.1) allow their replacement, as explained above, with P_l and C_S, respectively. This leads to slow and limited absorption (F <<0.90). Therefore, this slow absorption can be approximated with a constant rate of penetration:

$$(\text{Rate of penetration})_{IV} = P_l \cdot (SA)_i (C_S) = k_{IV} = \frac{F_i D}{\tau_i} \tag{3.16}$$

where k_{IV} denotes the constant penetration rate (mass/time units) for Class IV drugs, Fig. 3.2. Using the same syllogism delineated above, the differential equation describing the change of drug blood concentration C during the passage of drug from the stomach and small intestine (roughly, 4.86 h) [3] is as follows:

$$V_d \frac{dC}{dt} = k_{IV} - k_{el} C V_d \tag{3.17}$$

The passage of Class IV drugs to the colon via the ileocecal valve can either result in the termination of drug absorption or the significant reduction of the rate of drug penetration since the effective surface area is much smaller in the colon $(SA)_c$ and the amount of unabsorbed drug at the ileocecal valve is equal to $(1 - F_i)D$:

$$(\text{Rate of penetration})_{IV,c} = P_l \cdot (SA)_c (C_S) = k_{IV,c} = \frac{(1 - F_i)D}{\tau_c - \tau_i} \tag{3.18}$$

where $k_{IV,c}$ denotes the constant penetration rate (mass/time units) for Class IV drugs in the colon. Accordingly, the change of drug blood concentration C as a function of time assuming one-compartment model disposition for Class IV drugs in the colon is:

$$V_{\rm d}\frac{{\rm d}C}{{\rm d}t} = k_{\rm IV,c} - k_{\rm el}CV_{\rm d} \tag{3.19}$$

As explained above, this equation roughly holds from 4.86 h to the time needed for the drug to reach the non-absorptive sites of the colon, $\tau_{\rm c}$, (< 20.28 or < 31.95 h), i.e., the colon transit time for a single- or multi-unit formulation, respectively [3]. Beyond, this time point, $\tau_{\rm c}$, the drug is only eliminated from the body. Hence, the drug concentration decreases according to Eq. (3.11) for $t \geq \tau_{\rm c}$.

This theoretical section of oral drug absorption was based on (i) the finite absorption time concept, (ii) the physiologically based transit times reported in the literature [3], and (iii) the basic drug properties, namely, solubility and permeability, which have been adopted by the regulatory authorities as the key factors controlling oral drug absorption [5, 6]. However, the reader should be aware of the qualitative character of biopharmaceutics classification system, which implies large differences in the drug properties among the drugs of the same Class. Accordingly, the theoretical aspects developed here can be considered as a general framework of drug absorption, while the in vivo drug behavior can vary remarkably even for drugs of the same Class [5, 6]. Moreover, deviations from the general modeling framework may be applied in accordance with the experimental observations. For example, a drug may exhibit a regional rate of absorption differences in the various segments of the small intestines, e.g., jejunum and ileum. In such a case, two successive constant input rates can be considered. Although the development of models was based on one-compartment model disposition, similar equations can be written assuming two-compartment model disposition. For the purposes of the present work, we quote below the relevant equations assuming no absorption from the colon:

$$V_d\frac{{\rm d}C}{{\rm d}t} = k_{01} - (k_{12} + k_{10})C + k_{12} \qquad \text{for } 0 < t \leq \tau_{\rm i} \tag{3.20}$$

$$\frac{{\rm d}C}{{\rm d}t} = -(k_{12} + k_{10})C + k_{21}C_P \qquad \text{for } t > \tau_{\rm i} \tag{3.21}$$

where k_{12}, k_{10}, and k_{21} are the microconstants of the two-compartment model, $C_{\rm P}$ is the drug concentration in the peripheral compartment, and k_{01} can either be $k_{\rm I}$ or $k_{\rm II}$ or $k_{\rm III}$ or $k_{\rm IV}$ depending on the drug class type.

3.3 Simulations

Figure 3.3 shows examples of Class I drugs with rapid (Fig. 3.3a) and less rapid (Fig. 3.3b) absorption. The two input rates are 133.33 and 50.0 mg/h, respectively. Figure 3.3c shows a very rapid drug absorption (input rate = 1000 mg/h) in comparison with a set of curves generated from the Bateman equation (Eq. 1.2) using different values for the absorption rate constant, while in all cases a common value (0.1 h^{-1}) for the elimination rate constant is used. Visual inspection indicates

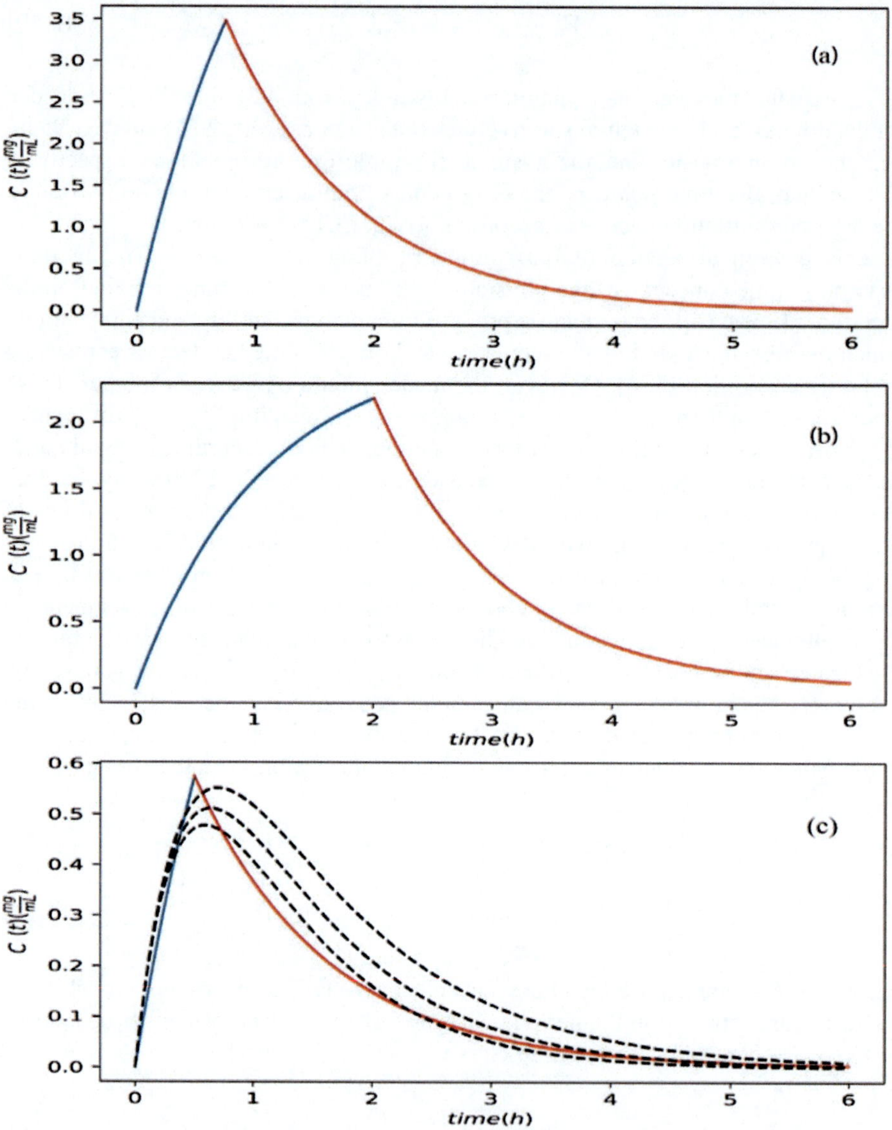

Fig. 3.3 (C, t) profiles for Class I drugs. Key: (**a**) Rapid drug absorption generated from Eqs. (3.4) and (3.6) adjusting $D = 100$ mg, $\tau_i = 0.75$ h, $V_d = 100$ L, $k_{el} = 0.1$ h^{-1} (**b**) Less rapid absorption compared to (**a**) considering $D = 100$ mg, $\tau_i = 2$ h, $V_d = 100$ L, $k_{el} = 0.1$ h^{-1} (**c**) Very rapid drug absorption (continuous line) considering $D = 500$ mg, $\tau_i = 0.5$ h, $V_d = 680$ L, $k_{el} = 0.1$ h^{-1}. The three dashed curves are generated from Eq. (1.2) (Bateman equation) using $D = 500$ mg, $V_d = 680$ L all sharing a common absorption rate constant $k_a = 2.23$ h^{-1}, while k_{el} values from top to bottom are 0.85, 1.05, 1.25 h^{-1}, respectively. The blue lines correspond to drug absorption and the red lines represent the drug's elimination phase

that the plots exhibit only minor differences around the peak of the curves; accordingly, the justification of the operating model under in vivo conditions will be rather impossible since the fitted curves will be indistinguishable when data with experimental error are used.

Figure 3.4a shows examples of Class II or III or IV drugs assuming various input rates for drug absorption in the colon lasting 6 h while a common input rate 2.1 mg/h in the stomach/small intestine lasting 4 h has been used. The blue curve represents the most common case since no absorption is taking place in the colon. The green and purple curves exhibit a monotonic decrease of drug concentration until time τ_c since the input rates in the colon (0.5 mg/h and 1 mg/h, respectively) are much smaller than the input rate in the stomach or small intestine, 2.05 mg/h. The brown curve exhibits a sustained type of profile since the drug's absorption in the colon (1.8 mg/h) has not been reduced drastically in comparison with the input rate (2.05 mg/h) in the stomach/small intestine. The orange curve represents the continuation of the drug's absorption in the colon for 6 more hours with an imperishable input rate (2.1 mg/h). Figure 3.4b is a representative example of a drug exhibiting regional permeability differences in the small intestine. Three successive input rates of 1.4, 1, and 0.8 mg/h in jejunum, ileum, and colon, respectively, have been simulated; the corresponding duration of the processes are 2, 1, and 7 h, respectively.

3.3.1 Physiological Considerations

The physiological aspects of the models based on the FAT concept rely on the physiological/anatomical differences between the two regions, small intestine and large intestine. It is widely known today that because of its permeability, large surface area, and high blood flow, the small intestine is the primary site for drug absorption, Fig. 3.2. In fact, a monolayer of enterocytes that is characterized by protrusions that extend into the gut lumen, called villi, result in a potential absorptive surface area of 60 m^2 in both the jejunum and the ileum [7]. On the contrary, the colon surface area totals around 0.25 m^2 as there are no villi [8]. This huge anatomical difference causes a very large difference in the rate of drug absorption, Eqs. (3.2), (3.7), (3.9), (3.12), and (3.14). Besides, the drug's transport from the gastrointestinal lumen to the portal vein relies on the sink conditions' principle of the universally accepted first-order drug absorption notion. This is substantiated by the fact that the blood in the portal vein has a velocity of 20–40 cm/s [4], which does not allow Fick's reversibility considerations for the drug transfer. In parallel, the small intestine was presented (Fig. 3.2) as a homogeneous compartment in terms of drug uptake. However, drug absorption can be higher from the jejunum than the ileum. A relevant simulation is shown in Fig. 3.4b where the initial absorption rate from the jejunum is 1.4 mg/h and lasts 2 h followed by an absorption rate of 1.0 mg/h from the ileum for 1 h, while the rate of drug absorption in the colon is equal to 0.8 mg/h and lasts 7 h.

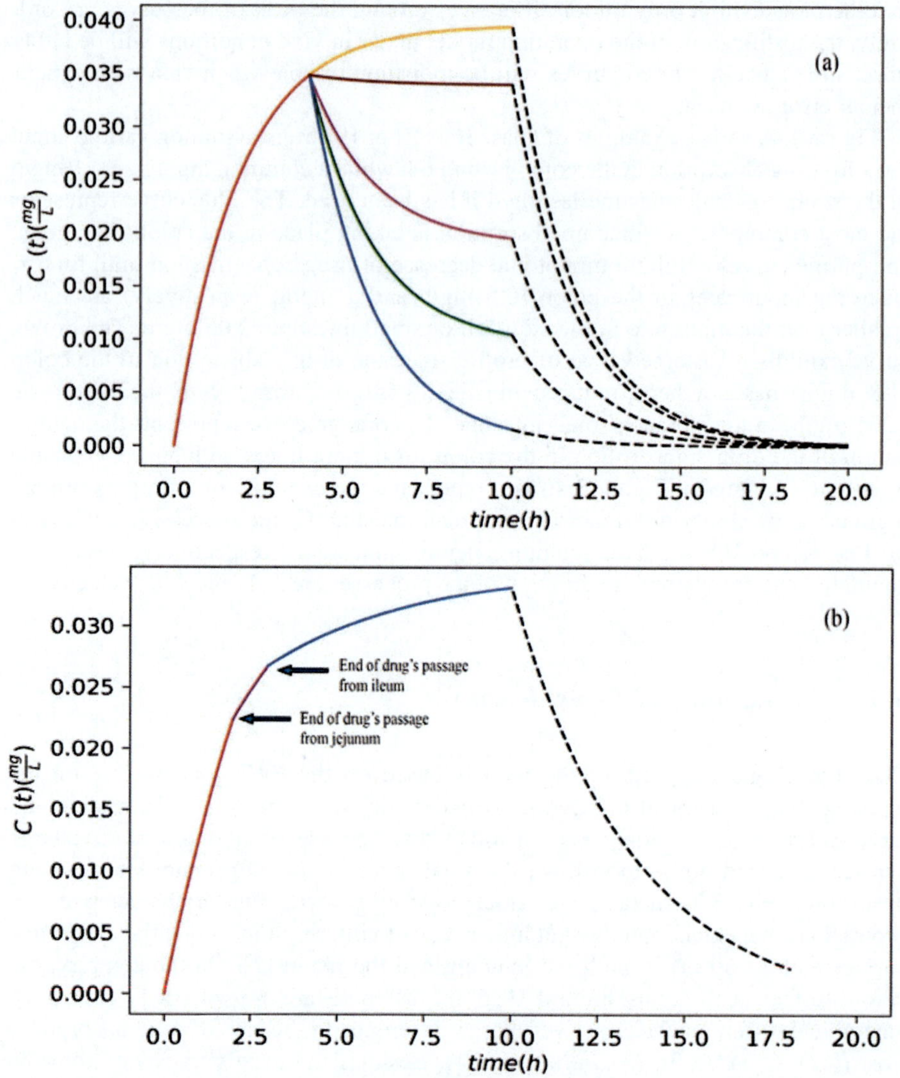

Fig. 3.4 (**a**) (C, t) curves for various drugs (Class II, or III, or IV) assuming a common (i) elimination rate constant, $k_{\mathrm{el}} = 0.53\ \mathrm{h}^{-1}$ (ii) small intestine input rate 2.1 mg/h, and (iii) volume of drug distribution 100 L. All drugs are absorbed in the stomach/small intestine, within 4 h. After that point, absorption in the colon, which lasts 6 h, is observed for all colored lines except for the blue line. Key: $k_{\mathrm{II,c}}$, or $k_{\mathrm{III,c}}$, or $k_{\mathrm{IV,c}}$ 2.05, 1.80, 1.00, 0.50, 0 mg/h from top to bottom, respectively (**b**) (C, t) curve for a drug exhibiting three successive constant rates of absorption 1.4, 1.0, 0.8 mg/h in the jejunum, ileum and colon, respectively. The duration of absorption is 2, 1, and 7 h, respectively. The arrows indicate the termination of absorption at the end of the two segments of the small intestine. The drug dose is equal to 100 mg, $k_{\mathrm{el}} = 0.23\ \mathrm{h}^{-1}$, while the volume of drug distribution is equal to 100 L

The unique features of the models based on the FAT concept are the finite termination times for the absorption phases, τ_i and τ_c, respectively. The upper limit for τ_i is 5 h, Fig. 3.2, with the most frequently observed values in the literature in the range 1–3 h depending on the drug's biopharmaceutical properties. The upper limit for τ_c is 30 h, Fig. 3.2, while the most usual values for τ_c are unknown since estimates for τ_c have not been explored so far. However, a large number of in vivo studies based on imaging techniques like gamma scintigraphy or magnetic resonance imaging coupled with drug blood measurements have shown that the completion of the absorption phase is terminated during the drug's passage from the small intestine, e.g., erythromycin study [9]. Needless to say that according to the current theory (Eq. 1.2, Bateman equation), the termination of either the elimination or the absorption phase is irreconcilable [1, 10–12].

3.3.2 Drug Absorption: Reconsidered in Terms of the FAT

The simulations of Fig. 3.3 lead to important observations for Class I drugs. First of all, there are conceptual differences in the concentration, time maxima generated from the models based on the FAT concept, and the corresponding parameters adhering to the Bateman equation. For Eq. (1.2), C_{max} corresponds to the steady state concentration derived from the classical equality (Rate in = Rate out) [13]; this applies to t_{max} too [13]. On the contrary, the pair ($C(\tau_i)$, τ_i) denotes the end of the drug's absorption phase, i.e., for a Class I drug, no more drug is available for absorption. The value of τ_i for Class I drugs reflect the high values of the biopharmaceutical properties solubility and permeability; in theory, the smaller the value of τ_i, the higher "the Class I character" of drug. Besides, the initial absorption phase can be almost linear (Fig. 3.3a) or non-linear (Fig. 3.3b) depending on the magnitude of the elimination rate constant in comparison with the magnitude of the input rate. In parallel, the declining limbs for all models based on the FAT concept concave upward and represent a single elimination phase. On the contrary, the declining limb of the curves generated from Eq. (1.2) concave downward initially and then upward beyond the inflection point at $2t_{max}$ [1]. However, the similarity of the curves is remarkable. For drug Classes II, III, and IV, drug absorption beyond the small intestine can be considered. During the time period τ_i–τ_c, the identical form of Eqs. (3.10), (3.15), and (3.19) for drug Classes II, III, and IV, respectively, allow a common consideration. The (C, t) profiles during the time interval τ_i–τ_c is heavily dependent on the relative magnitude of the two terms at the right-hand side of Eqs. (3.10), (3.15), and (3.19). Obviously, during the time interval τ_i–τ_c, the derivative of Eqs. (3.10), (3.15), and (3.19) is constantly negative for the three monotonically decreasing profiles at the bottom of Fig. 3.4a. The bottom curve of Fig. 3.4a represents the most common case whereas drug absorption is not observed in the colon and τ_c becomes meaningless since drug absorption is terminated in the small intestine ($\leq \tau_i$). For the top curve of Fig. 3.4, the derivative of Eqs. (3.10), (3.15), and (3.19) is initially positive, becomes equal to zero at pseudo-steady state and then negative beyond time τ_c (10 h in the example of Fig. 3.4a). For all examples depicted in Fig. 3.4a, a common elimination phase (generated from Eq. 3.6, $t \geq \tau_c$) is shown. It can be seen that a single elimination phase prevails during the time course of drug in the body beyond τ_i only for the bottom curve. Besides, the separation of the elimination phase cannot be easily accomplished under in vivo conditions in several cases, e.g., the second from the bottom curve of Fig. 3.4a. Moreover, multiple regional input rates can be considered if a prolonged absorption phase is observed, Fig. 3.4b. Overall, this analysis reveals that rich kinetics can be encountered for Class II, III, and IV drugs, which is dependent on the rate of drug absorption in the colon.

3.4 Fits to Data

Methodology All equations were integrated using Python's open-source SciPy v.1.4.1 ecosystem. Its embedded solvers using Adams' method for non-stiff problems provided us the continuous solutions of the first-order differential equations. Data fitting was performed for four different drugs obtained from literature, namely, cephradine [14], ibuprofen [14], flurbiprofen [15], and itraconazole [16]. The models based on the FAT concept were fitted to the data using SciPy's curve_fit function. The function uses the Levenberg–Marquardt algorithm which has become a standard technique for non-linear least-squares problems [17]. The generated plots depict two discrete curves united at the top instead of a smooth curve. The choice of binding the proposed pairwise equations with a Heaviside function [18] in order to erase discontinuities at C_{max} was also tested. Although it provided very competent R^2 metrics and almost identical fittings, the presentation of two discrete curves that join with a small discontinuity at their top, attributed to a small extrapolation at this data point due to the fitting procedure, was preferred. The selection of this format was navigated by the need to underline the two distinct physiological absorption and elimination processes. All corresponding plots were illustrated with SciPy's matplotlib.

Data Analysis All (C,t) data belonging to the declining limbs of cephradine, ibuprofen, and flurbiprofen profiles were analyzed using semi-logarithmic plots, Fig. 3.5. For all sets of data very nice fittings were obtained with R^2 values 0.993, 0.989, and 0.980 while the elimination rate constants estimates were 0.97, 0.33, and 0.11 h^{-1} for cephradine, ibuprofen, and flurbiprofen, respectively [2].

The estimates of the respective elimination rate constants for cephradine, ibuprofen, and flurbiprophen were coupled with Eq. (3.8) and fitted to the entire set of data, Fig. 3.6a–c. The R^2 values of the fittings were 0.9723, 0.9961, and 0.9908, while the estimates for optimal k_{II} were 15.07, 11.12 mg/h for ibuprofen and flurbiprophen, respectively. A meaningless estimate was derived for cephradine input rate upon curve fitting. This is attributed to the very steep change of cephradine concentration

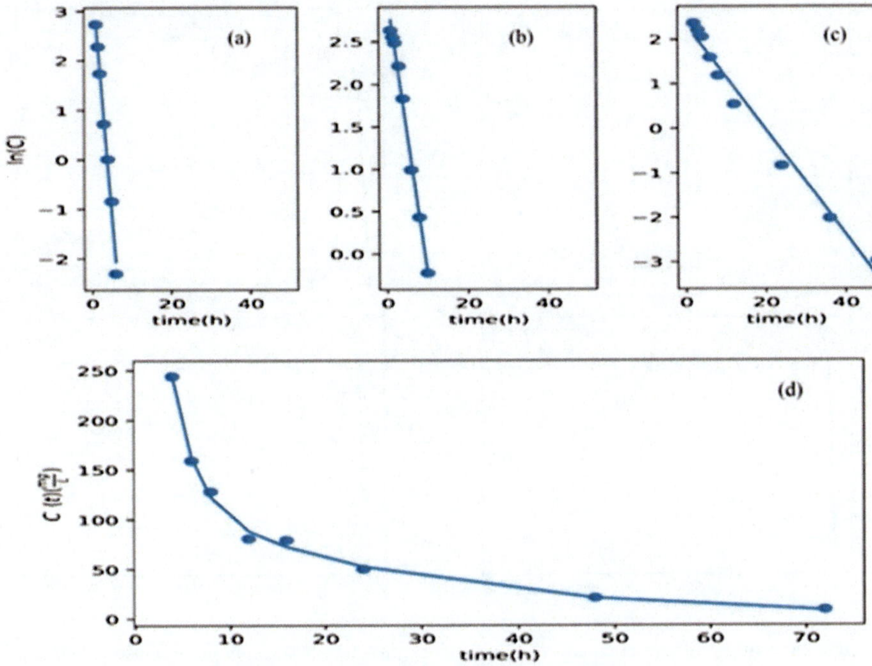

Fig. 3.5 Computation of the slope of one-compartment model drugs [(**a**), (**b**), (**c**)] using semi-logarithmic plots for all data points of the declining limb of the $(C,\, t)$ curve [2]. Key: (**a**) Cephradine, Dose = 500 mg, $k_{\mathrm{el}} = 0.97\ \mathrm{h}^{-1}$, $R^2 = 0.991$; (**b**) Ibuprofen, Dose = 200 mg, $k_{\mathrm{el}} = 0.33\ \mathrm{h}^{-1}$, $R^2 = 0.993$; (**c**) Flurbiprofen, Dose = 100 mg, $k_{\mathrm{el}} = 0.11\ \mathrm{h}^{-1}$, $R^2 = 0.980$. (**d**) Non-linear two-compartment model fit to all itraconazole data points of the declining limb of the $(C,\, t)$ curve of a dose 200 mg. Computation of optimal parameters: $A = 679.46$ mg/mL, $\alpha = 0.415\ \mathrm{h}^{-1}$, $B = 132.68$ mg/mL, $\beta = 0.038\ \mathrm{h}^{-1}$ from all elimination phase data points, $R^2 = 0.9985$

in the absorption phase. A graphical estimate of 14.81 mg/h was derived from the data points assuming a linear change.

Since itraconazole [16] is a two-compartment model, Class II drug the entire set of the declining limb of concentration–time data was analyzed using the general equation that describes a two-compartment disposition model: $C = Ae^{-\alpha t} + Be^{-\beta t}$ [13]. The estimates for the constants A, B, α, β were computed with a non-linear least-squares approach using the Levenberg–Marquardt algorithm [18]. A nice fit was obtained, $R^2 = 0.9768$, Fig. 3.6d; the estimates for A, B, α, β were 679.46 mg/L, 13.26 mg/L, $0.41\ \mathrm{h}^{-1}$, $0.038\ \mathrm{h}^{-1}$, respectively. Subsequently, estimates for the microconstants, k_{12}, k_{10}, k_{21} were derived algebraically [13] and used in a curve-fitting exercise of Eqs. (3.20) and (3.21) to the entire set of itraconazole experimental data, Fig. 3.6d. The algorithm converged with an R^2 metric 0.9797, while the estimate for k_{II} was 990 mg/h. The analysis of all data demonstrates that the absorption of drugs has been terminated at 1.0, 2.0, 0.7, 3.0 h for cephradine, ibuprofen, flurbiprofen, and

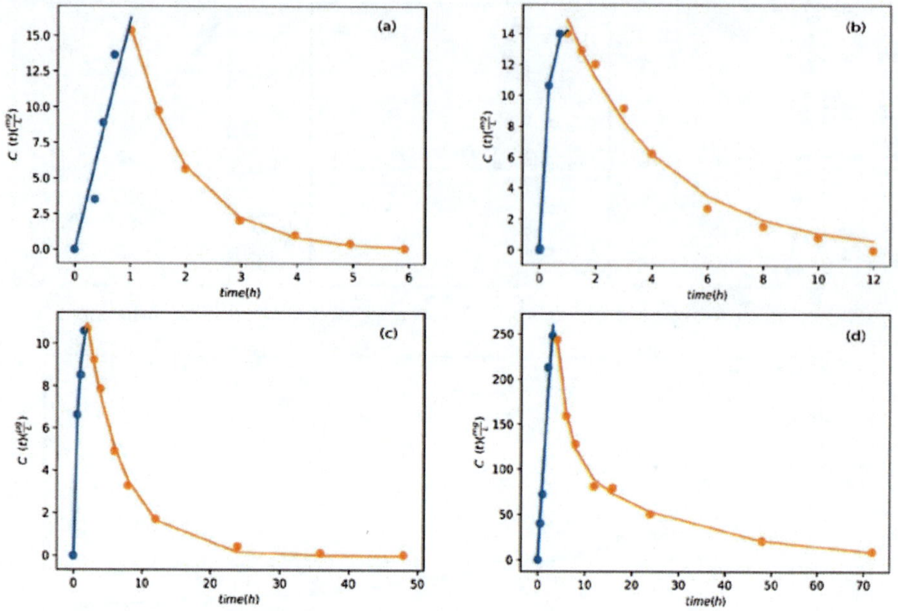

Fig. 3.6 Curve fitting of model equations to all data points. Key: (**a**) Cephradine. Dose = 500 mg, $R^2 = 0.9723$. (**b**) Ibuprofen, Dose = 200 mg, $R^2 = 0.9961$. (**c**) Flurbiprofen, Dose = 100 mg, $R^2 = 0.9908$. (**d**) Itraconazole, Dose = 200 mg, $R^2 = 0.9797$. The blue lines correspond to drug absorption and the orange lines represent the drug's elimination phase

itraconazole, respectively, Fig. 3.6 [2]. In other words, all drugs were absorbed in the small intestine, which is in accord with the general and common wisdom belief.

3.5 Epilogue

The finite absorption time (FAT) concept has been sporadically used in the past [19–21] in the context of classical first-order absorption kinetics, i.e., Eq. (1.2) applies only for a specific time period. However, the blood flow in the portal vein has a velocity of 20–40 cm/s [4] ensuring sink conditions for the drug transfer from the gastrointestinal lumen to the bloodstream. Therefore, the passive absorption of drugs follows zero-order kinetics as the result of the dynamics of blood flow in the microenvironment of gastrointestinal membrane/vena cava region. This chapter describes the first formal development of the FAT concept using a minimal physiological model, Fig. 3.2. In this first application of the models based on the FAT concept, the absorption and elimination phase data were analyzed separately. The formal definition and use of relevant Physiologically Based Finite Time Pharmacokinetic (PBFTPK) models for the analysis of blood concentration, time data using the entire set of experimental data are described in Chap. 4.

References

1. Macheras P (2019) On an unphysical hypothesis of Bateman equation and its implications for pharmacokinetics. Pharm Res 36:94. https://doi.org/10.1007/s11095-019-2633-4
2. Macheras P, Chryssafidis P (2020) Revising pharmacokinetics of oral drug absorption: I Models based on biopharmaceutical/physiological and finite absorption time concepts. Pharm Res 37: 187. https://doi.org/10.1007/s11095-020-02894-w. (Erratum, Pharm. Res. 37, 206 (2020) https://doi.org/10.1007/s11095-020-02935-4)
3. Abuhelwa A, Foster D, Upton R (2016) A quantitative review and meta-models of the variability and factors affecting Oral drug absorption-part II: gastrointestinal transit time. AAPS J 18:1322–1333. https://doi.org/10.1208/s12248-20016-209953-207
4. Iranpour P, Lall C, Houshyar R, Helmy M, Yang A, Choi JI, Ward G, Goodwin SC (2016) Altered Doppler flow patterns in cirrhosis patients: an overview. Ultrasonography 35:3–12. https://doi.org/10.14366/usg.15020
5. Food and Drug Administration (2017) Center for Drug Evaluation and Research (CDER). Waiver of In Vivo Bioavailability and Bioequivalence Studies for Immediate-Release Solid Oral Dosage Forms Based on a Biopharmaceutics Classification System. Guidance for Industry U.S. Department of Health and Human Services
6. European Medicines Agency (2017) Committee for medicinal products for human use (CHMP) Guideline on the investigation of bioequivalence. London
7. Vertzoni M, Augustijns P, Grimm M, Koziolek M, Lemmens G, Parrott N, Pentafragka C, Reppas C, Rubbens J, Van Den Abeele J, Vanuytsel T, Weitschies W, Wilson CG (2019) Impact of regional differences along the gastrointestinal tract of healthy adults on oral drug absorption: an UNGAP review. Eur J Pharm Sci 134:153–175. https://doi.org/10.1016/j.ejps.2019.04.013
8. Kararli T, Searle GD (1995) Comparison of the gastrointestinal anatomy, physiology, and biochemistry of humans and commonly used laboratory animals. Biopharm Drug Disp 16: 351–380. https://doi.org/10.1002/bdd.2510160502
9. Digenis GA, Sandefer EP, Parr AF, Beihn R, McClain C, Scheinthal BM, Ghebre-Sellassie I, Iyer U, Nesbitt RU, Randinitis E (1990) Gastrointestinal behavior of orally administered radiolabeled erythromycin pellets in man as determined by gamma scintigraphy. J Clin Pharmacol 30:621–631. https://doi.org/10.1002/j.1552-4604.1990.tb01865.x
10. Dost FH (1953) Der Blutspiegel. Kinetik der Konzentrationsabläufe in der Kreislaufflüssigkeit. Thieme, Leipzig
11. Bateman H (1910) The solution of a system of differential equations occurring in the theory of radioactive transformations. Proc Camb Philos Soc Math Phys Sci 15:423–427. https://www.biodiversitylibrary.org/item/97262#page/487/mode/1up or https://archive.org/details/cbarchive_122715_solutionofasystemofdifferentia1843
12. Garrett ER (1994) The Bateman function revisited: a critical reevaluation of the quantitative expressions to characterize concentrations in the one compartment body model as a function of time with first-order invasion and first-order elimination. J Pharmacokin Biopharm 22:103–128. https://doi.org/10.1007/BF02353538
13. Niazi S (1979) Textbook of biopharmaceutics and clinical pharmacokinetics. Appleton-Century-Crofts, New York
14. Chen ML (1992) An alternative approach for assessment of rate of absorption in bioequivalence studies. Pharm Res 9:1380. https://doi.org/10.1023/A:1015842425553
15. Szpunar GJ, Albert KS, Bole GG, Dreyfus JN, Lockwood GF, Wagner JG (1987) Pharmacokinetics of flurbiprofen in man I. Area/dose relationships. Biopharm Drug Disp 8:273–228. https://doi.org/10.1002/bdd.2510080308
16. Hardin TC, Graybill JR, Fetchick R, Woestenborghs R, Rinaldi MG, Kuhn KG (1988) Pharmacokinetics of itraconazole following oral administration to normal volunteers. Antimicrobial Agents Chemother 32:1310–1313. https://doi.org/10.1128/AAC.32.9.1310

17. Arfken GB, Weber HJ, Harris FE (2013) Mathematical methods for physicists, 7th edn. Elsevier, Waltham
18. Press WH, Teukolsky SA, Vetterling WT, Flannery BP (2007) Numerical recipes 3rd edition: the art of scientific computing. Cambridge University Press
19. Lovering EG, McGilveray IJ, McMillan I, Tostowaryk W (1975) Comparative bioavailabilities from truncated blood level curves. J Pharm Sci 64:1521–1524. https://doi.org/10.1002/jps.2600640921
20. Sugano K (2012) Biopharmaceutics modeling and simulations: theory, practice, methods, and applications. Wiley, Hoboken
21. Sugano K (2021) Lost in modelling and simulation? ADMET DMPK 9:75–109. https://doi.org/10.5599/admet.923

Chapter 4
The Rise of Physiologically Based Finite Time Pharmacokinetic (PBFTPK) Models

All models are wrong, but some are useful.
George E. P. Box (1919–2013)

Contents

Abstract Differential equations are written describing zero-order absorption in one- and two-compartment models, which are solved analytically. An assortment of possible pharmacokinetic time profiles shows the potential of these models through detailed simulations. A variety of experimental data is then used to determine the best models and optimum parameter values for the duration of the drug's absorption and other kinetic rate constants. A new drug biopharmaceutic–pharmacokinetic classification system is proposed based on the magnitude of the finite absorption time and the elimination rate constant.

Keywords Oral pharmacokinetics · Physiologically based finite time pharmacokinetic models · Finite absorption time · Oral drug absorption · Almotriptan · Cyclosporine · Ibuprofen · Niraparib · Paracetamol

4.1 Models

The concept of Finite Absorption Time (FAT) developed in Chap. 3 implies that delivery of a drug from the GI tract to the blood follows simple, zero-order, kinetics, and ceases after time τ. Given the sink conditions prevailing in the absorption process due to the high blood flow rate in the vena cava [1], Fig. 4.1, the rate of

P. Macheras, A. A. Tsekouras, *Revising Oral Pharmacokinetics, Bioavailability and Bioequivalence Based on the Finite Absorption Time Concept*,
https://doi.org/10.1007/978-3-031-20025-0_4

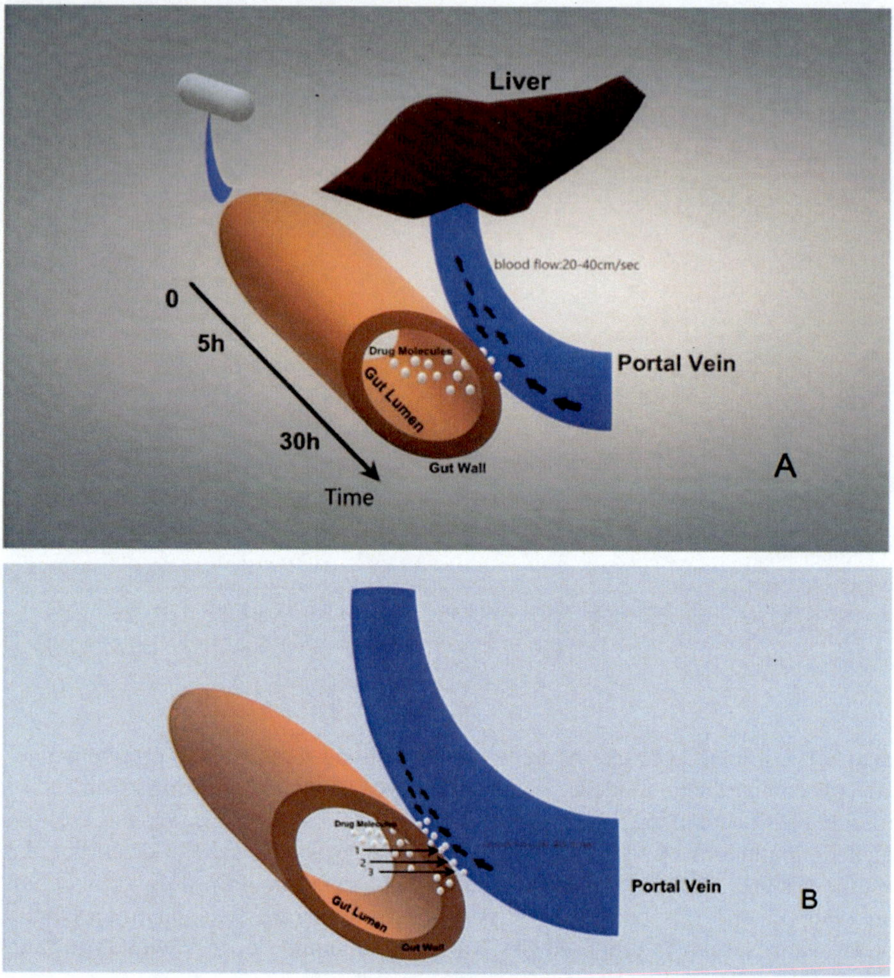

Fig. 4.1 (**a**) Schematic of the passive transfer of dissolved drug molecules (white spheres) from the gut lumen to portal vein. The blood flow in the portal vein, 20–40 cm/s ensures sink conditions for the passive drug transfer due to its continuous removal from the portal vein to liver. The physiological time limits of 5 and 30 h for drug absorption from the small intestines and colon, respectively, are shown on the time axis. (**b**) Enlargement of the region gut wall-portal vein for the drug transfer; the arrows indicate up to three successive constant input rates for the dissolved drug molecules (white spheres) passive transfer under sink conditions

change of drug concentration, C, in the bloodstream ignoring disposition/elimination processes: should obey the differential equation:

$$\frac{\mathrm{d}C}{\mathrm{d}t} = \frac{FD}{\tau V_{\mathrm{d}}} \tag{4.1}$$

for a zero-order absorption process that lasts for time τ. This chapter describes the development and use of the Physiologically Based Finite Time Pharmacokinetic (PBFTPK) [1] models, which are the logical consequence of the FAT concept developed in Chap. 3.

Due to the anatomical–physiological characteristics of the gastrointestinal tract, drugs with different biopharmaceutical properties, e.g., solubility, permeability, and ionization, can exhibit one or two or three successive constant input rates, Fig. 4.1b. It was shown in Chap. 3 that although the physical properties may vary and hence a drug may belong to any of the four BCS classes, the mathematical expressions, i.e., the differential equations, have the same form. Consequently, their solutions do not reflect the BCS classification they fall under.

Depending on whether the model considered involves one ($m = 1$) or two ($m = 2$) or more ($m > 2$) compartments and also assuming a single ($p = 1$) or multiple ($p \geq 2$) successive absorption stages, the following Physiologically Based Finite Time Pharmacokinetic (PBFTPK) termed p-PBFTPK-m can be formulated, Fig. 4.2. For the metabolized drugs following non-linear Michaelis–Menten disposition kinetics we coin the term p-PBFTPK-m(MM). A schematic representation of all models exhibiting linear or non-linear disposition kinetics is shown in Fig. 4.2 [1].

The full differential equations for the linear models are listed in Table 4.1. $C(t)$ is the drug concentration in the blood (the only or primary compartment) and $C_{\mathrm{P}}(t)$ the concentration in peripheral compartment. For simplicity's sake, both volumes V_{c} and V_{p} corresponding to the central and peripheral compartments are taken equal, since actual values for $C_{\mathrm{P}}(t)$ cannot be monitored and would only differ by a constant scaling factor equal to the ratio of the volumes.

The solutions to the differential equations of Table 4.1 corresponding to the one- and two-compartment models are given in Tables 4.2, 4.3, 4.4, and 4.5. The forms given in Tables 4.2 and 4.3 are not explicit but are compact and highly amenable to numerical calculations. The final forms of the same expressions are given in Tables 4.4 and 4.5. The observing reader will note that Eqs. (4.80)–(4.83) are identical to Eqs. (4.52)–(4.55). The final substitutions were avoided here because of the large size of the resulting expressions.

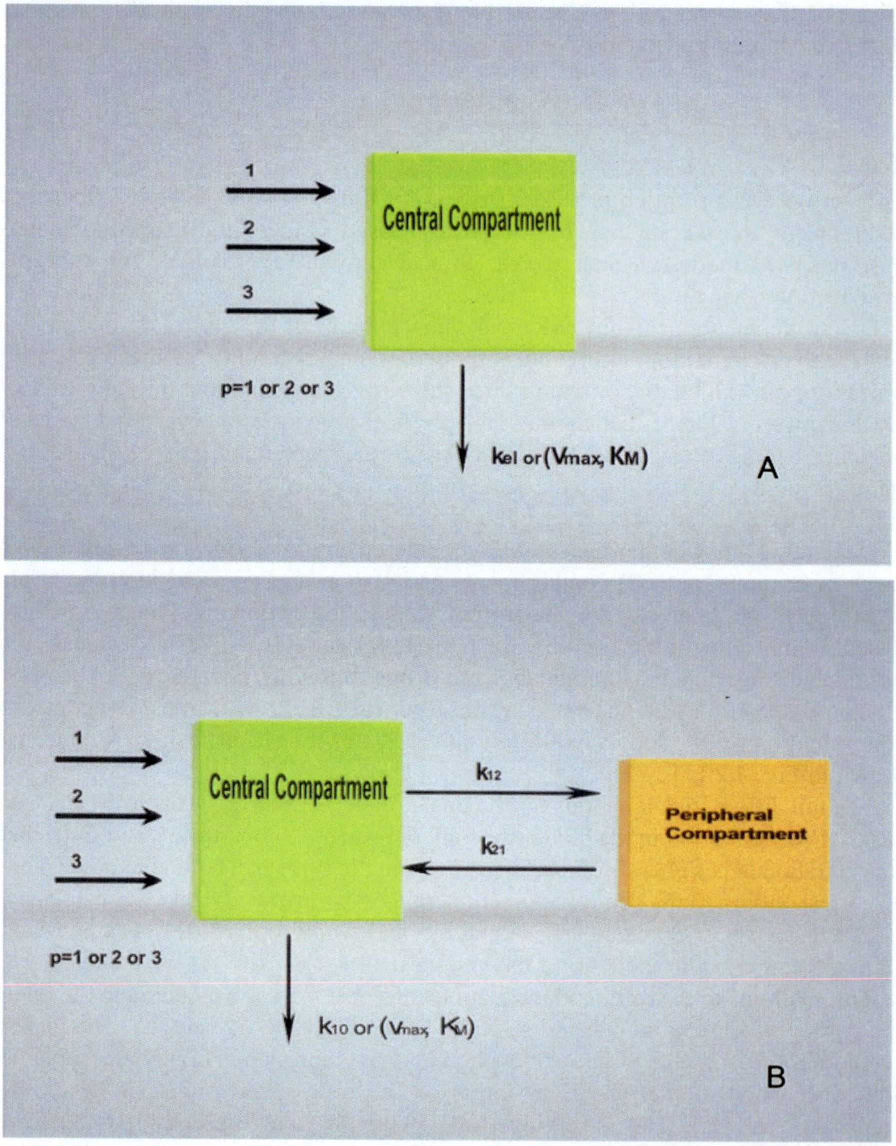

Fig. 4.2 Schematic representation of one-compartment (**a**) and two-compartment (**b**) *p*-PBFTPK-*m* models. In all cases, the horizontal arrows at the left-hand side of the central compartment denote the number of successive constant drug input rates, not necessarily of the same drug amount or duration; k_{el} is the elimination rate constant, k_{10} is the elimination rate constant of the central compartment of the two-compartment model drugs; k_{12} and k_{21} are the disposition micro-constants for the transfer of drug from the central to peripheral compartment and vice versa, respectively; V_{max} and K_M correspond to the maximum biotransformation rate and the constant of the Michaelis–Menten kinetics

Table 4.1 Differential equations

p	m	Kinetic (differential) equations	$t_1 < t \leq t_2$		Eq.
1	1	$\frac{dC}{dt} = \frac{FD}{\tau V_d} - k_{el}C$	0	τ	(4.2)
		$\frac{dC}{dt} = -k_{el}C$	τ	∞	(4.3)
2	1	$\frac{dC}{dt} = \frac{F_1 D}{\tau_1 V_d} - k_{el}C$	0	τ_1	(4.4)
		$\frac{dC}{dt} = \frac{F_2 D}{\tau_2 V_d} - k_{el}C$	τ_1	$\tau_1 + \tau_2$	(4.5)
		$\frac{dC}{dt} = -k_{el}C$	$\tau_1 + \tau_2$	∞	(4.6)
3	1	$\frac{dC}{dt} = \frac{F_1 D}{\tau_1 V_d} - k_{el}C$	0	τ_1	(4.7)
		$\frac{dC}{dt} = \frac{F_2 D}{\tau_2 V_d} - k_{el}C$	τ_1	$\tau_1 + \tau_2$	(4.8)
		$\frac{dC}{dt} = \frac{F_3 D}{\tau_3 V_d} - k_{el}C$	$\tau_1 + \tau_2$	$\tau_1 + \tau_2 + \tau_3$	(4.9)
		$\frac{dC}{dt} = -k_{el}C$	$\tau_1 + \tau_2 + \tau_3$	∞	(4.10)
1	2	$\frac{dC}{dt} = \frac{FD}{\tau V_c} - k_{12}C - k_{10}C + k_{21}C_P$	0	τ	(4.11)
		$\frac{dC_P}{dt} = k_{12}C - k_{21}C_P$			(4.12)
		$\frac{dC}{dt} = -k_{12}C - k_{10}C + k_{21}C_P$	τ	∞	(4.13)
		$\frac{dC_P}{dt} = k_{12}C - k_{21}C_P$			(4.14)
2	2	$\frac{dC}{dt} = \frac{F_1 D}{\tau_1 V_c} - k_{12}C - k_{10}C + k_{21}C_P$	0	τ_1	(4.15)
		$\frac{dC_P}{dt} = k_{12}C - k_{21}C_P$			(4.16)
		$\frac{dC}{dt} = \frac{F_2 D}{\tau_2 V_c} - k_{12}C - k_{10}C + k_{21}C_P$	τ_1	$\tau_1 + \tau_2$	(4.17)
		$\frac{dC_P}{dt} = k_{12}C - k_{21}C_P$			(4.18)
		$\frac{dC}{dt} = -k_{12}C - k_{10}C + k_{21}C_P$	$\tau_1 + \tau_2$	∞	(4.19)
		$\frac{dC_P}{dt} = k_{12}C - k_{21}C_P$			(4.20)
3	2	$\frac{dC}{dt} = \frac{F_1 D}{\tau_1 V_c} - k_{12}C - k_{10}C + k_{21}C_P$	0	τ_1	(4.21)
		$\frac{dC_P}{dt} = k_{12}C - k_{21}C_P$			(4.22)
		$\frac{dC}{dt} = \frac{F_2 D}{\tau_2 V_c} - k_{12}C - k_{10}C + k_{21}C_P$	τ_1	$\tau_1 + \tau_2$	(4.23)
		$\frac{dC_P}{dt} = k_{12}C - k_{21}C_P$			(4.24)
		$\frac{dC}{dt} = \frac{F_3 D}{\tau_3 V_c} - k_{12}C - k_{10}C + k_{21}C_P$	$\tau_1 + \tau_2$	$\tau_1 + \tau_2 + \tau_3$	(4.25)
		$\frac{dC_P}{dt} = k_{12}C - k_{21}C_P$			(4.26)
		$\frac{dC}{dt} = -k_{12}C - k_{10}C + k_{21}C_P$	$\tau_1 + \tau_2 + \tau_3$	∞	(4.27)
		$\frac{dC_P}{dt} = k_{12}C - k_{21}C_P$			(4.28)

Table 4.2 Solutions to the one-compartment model differential equations for 1, 2, or 3 successive input stages with recursive expressions

p	m	$C(t)$	$t_1 < t \leq t_2$		Eq.
1	1	$\frac{FD}{\tau V_d k_{el}}\left(1 - e^{-k_{el}t}\right)$	0	τ	(4.29)
		$C(\tau)e^{-k_{el}(t-\tau)}$	τ	∞	(4.30)
2	1	$\frac{F_1 D}{\tau_1 V_d k_{el}}\left(1 - e^{-k_{el}t}\right)$	0	τ_1	(4.31)
		$C(\tau_1)e^{-k_{el}(t-\tau_1)} + \frac{F_2 D}{\tau_2 V_d k_{el}}\left(1 - e^{-k_{el}(t-\tau_1)}\right)$	τ_1	$\tau_1 + \tau_2$	(4.32)
		$C(\tau_1 + \tau_2)e^{-k_{el}(t-\tau_1-\tau_2)}$	$\tau_1+\tau_2$	∞	(4.33)
3	1	$\frac{F_1 D}{\tau_1 V_d k_{el}}\left(1 - e^{-k_{el}t}\right)$	0	τ_1	(4.34)
		$C(\tau_1)e^{-k_{el}(t-\tau_1)} + \frac{F_2 D}{\tau_2 V_d k_{el}}\left(1 - e^{-k_{el}(t-\tau_1)}\right)$	τ_1	$\tau_1 + \tau_2$	(4.35)
		$C(\tau_1 + \tau_2)e^{-k_{el}(t-\tau_1-\tau_2)} + \frac{F_3 D}{\tau_3 V_d k_{el}}\left(1 - e^{-k_{el}(t-\tau_1-\tau_2)}\right)$	$\tau_1+\tau_2$	$\tau_1 + \tau_2 + \tau_3$	(4.36)
		$C(\tau_1 + \tau_2 + \tau_3)e^{-k_{el}(t-\tau_1-\tau_2-\tau_3)}$	$\tau_1+\tau_2+\tau_3$	∞	(4.37)

4.2 Simulations

Before we can use the analytic expressions derived in the previous section to parameterize pharmacokinetic data, we need to visualize these functions. Using Eqs. (4.57)–(4.84), we can simulate a large variety of pharmacokinetic profiles. Several such examples are provided below. Pertinent parameter values used are shown as an inset in every diagram. Additional information is also provided there regarding t_{max} so it can be contrasted with τ. For two-compartment models peripheral compartment concentrations $C_P(t)$ are also depicted for comparison (black broken line). In all plots, the solid triangle denotes the end of the absorption process(es).

In all the above plots, the $(C(\tau), \tau)$ pair is a discontinuity datum point. When $t_{max} = \tau$, there is a more patent change of the concentration-time curve in the neighborhood of the discontinuity time point, Figs. 4.3, 4.4a, b, 4.5b, 4.6, 4.7b and 4.8a, b, d. On the contrary, when $t_{max} < \tau$, the discontinuity datum point lies in the descending part of the concentration–time curve, Figs. 4.4c, 4.5a, c, 4.7a, and 4.8c, e, f, and therefore this change is less abrupt. In Fig. 4.9, one can see the change of the derivative dC/dt for two examples with $t_{max} = \tau$ and $t_{max} < \tau$. In the former case, the derivative changes from positive to negative values at $t_{max} = \tau$; in the latter case, the sign of the derivative is maintained negatively close to τ and throughout the descending portion of the curve. These plots demonstrate that under experimental conditions the estimation of τ will be easier when $t_{max} = \tau$. When $t_{max} < \tau$, the presence of experimental error and sparse sampling close to τ can make the estimation of τ impossible.

Table 4.3 Solutions to the two-compartment model differential equations for 1, 2, or 3 successive input stages with recursive expressions

p	m	$C(t)$, $C_P(t)$ †	$t_1 < t \le t_2$		Eq.
1	2	$C(t)=\frac{FD}{\tau V_c}\left(\frac{(k_{21}-\alpha)(1-e^{-\alpha t})}{a(\beta-\alpha)}+\frac{(k_{21}-\beta)\left(1-e^{-\beta t}\right)}{\beta(\alpha-\beta)}\right)$	0	τ	(4.38)
		$C_P(t)=\frac{FDk_{12}}{(\beta-\alpha)\tau V_c}\left[\frac{1}{\alpha}\left(1-e^{-\alpha t}\right)-\frac{1}{\beta}\left(1-e^{-\beta t}\right)\right]$			(4.39)
		$C(t)=\frac{FD}{\tau V_c}\frac{(k_{21}-\alpha)(1-e^{-\alpha\tau})}{a(\beta-\alpha)}e^{-\alpha(t-\tau)}+\frac{FD}{\tau V_c}\frac{(k_{21}-\beta)\left(1-e^{-\beta\tau}\right)}{\beta(\alpha-\beta)}e^{-\beta(t-\tau)}$	τ	∞	(4.40)
		$C_P(t)=\frac{k_{12}}{\beta-\alpha}\left(\left(C(\tau)+\frac{k_{21}P(\tau)}{k_{21}-\alpha}\right)e^{-\alpha(t-\tau)}-\left(C(\tau)+\frac{k_{21}C_P(\tau)}{k_{21}-\beta}\right)e^{-\beta(t-\tau)}\right)$			(4.41)
2	2	$C(t)=\frac{F_1D}{\tau_1V_c}\left[\frac{k_{21}-\alpha}{\alpha(\beta-\alpha)}\left(1-e^{-\alpha t}\right)+\frac{k_{21}-\beta}{\beta(\alpha-\beta)}\left(1-e^{-\beta t}\right)\right]$	0	τ_1	(4.42)
		$C_P(t)=\frac{F_1Dk_{12}}{(\beta-\alpha)\tau_1V_c}\left[\frac{1}{\alpha}\left(1-e^{-\alpha t}\right)-\frac{1}{\beta}\left(1-e^{-\beta t}\right)\right]$			(4.43)
		$C(t)=\frac{D}{(\beta-\alpha)^2V_c}\left(\frac{F_1}{\tau_1}\left(\left(1-\frac{k_{21}}{\beta}\right)\left(1-e^{-\beta\tau_1}\right)\right.\right.$ $-\left(1-\frac{k_{21}}{\alpha}\right)(1-e^{-\alpha\tau_1}))\left(\beta e^{-\beta(t-\tau_1)}-\alpha e^{-\alpha(t-\tau_1)}\right)+\frac{F_1}{\tau_1}k_{21}\left(\frac{\alpha-k_{10}}{\beta}\left(1-e^{-\beta\tau_1}\right)\right.$ $-\frac{\beta-k_{10}}{\alpha}(1-e^{-\alpha\tau_1}))\left(e^{-\beta(t-\tau_1)}-e^{-\alpha(t-\tau_1)}\right)$ $+\frac{F_2}{\tau_2}(\beta-\alpha)\left(\left(1-\frac{k_{21}}{\beta}\right)\left(1-e^{-\beta(t-\tau_1)}\right)-\left(1-\frac{k_{21}}{\alpha}\right)\left(1-e^{-\alpha(t-\tau_1)}\right)\right))$	τ_1	$\tau_1+\tau_2$	(4.44)
		$C_P(t)=\frac{k_{12}F_2D}{(\beta-\alpha)\tau_2V_c}\left[\frac{1}{\alpha}\left(1-e^{-a(t-\tau_1)}\right)-\frac{1}{\beta}\left(1-e^{-\beta(t-\tau_1)}\right)\right]$ $+\frac{k_{12}F_1D}{(\beta-\alpha)^2\tau_1V_c}\left[\left(k_{21}-\alpha+\frac{k_{12}k_{21}}{k_{21}-\alpha}\right)\frac{1-e^{-\alpha\tau_1}}{\alpha}e^{-a(t-\tau_1)}\right.$			(4.45)

(continued)

Table 4.3 (continued)

p	m	$C(t)$, $C_P(t)$ †	$t_1 < t \leq t_2$		Eq.
		$+\left(k_{21}-\beta+\frac{k_{12}k_{21}}{k_{21}-\beta}\right)\frac{1-e^{-\beta\tau_1}}{\beta}e^{-\beta(t-\tau_1)}]$			
		$C(t)=\frac{[C(\tau_1+\tau_2)(k_{21}-\alpha)+C_P(\tau_1+\tau_2)k_{21}]e^{-\alpha(t-\tau_1-\tau_2)}}{\beta-\alpha}+\frac{[C(\tau_1+\tau_2)(k_{21}-\beta)+C_P(\tau_1+\tau_2)k_{21}]e^{-\beta(t-\tau_1-\tau_2)}}{\alpha-\beta}$	$\tau_1+\tau_2$	∞	(4.46)
		$C_P(t)=\frac{k_{12}}{\beta-\alpha}\left(\left(C(\tau_1+\tau_2)+\frac{k_{21}C_P(\tau_1+\tau_2)}{k_{21}-\alpha}\right)e^{-\alpha(t-\tau_1-\tau_2)}-\left(C(\tau_1+\tau_2)+\frac{k_{21}C_P(\tau_1+\tau_2)}{k_{21}-\beta}\right)e^{-\beta(t-\tau_1-\tau_2)}\right)$			(4.47)
3	2	$C(t)=\frac{F_1D}{\tau_1V_c}\left[\frac{k_{21}-\alpha}{\alpha(\beta-\alpha)}(1-e^{-\alpha t})+\frac{k_{21}-\beta}{\beta(\alpha-\beta)}\left(1-e^{-\beta t}\right)\right]$	0	τ_1	(4.48)
		$C_P(t)=\frac{F_1Dk_{12}}{(\beta-\alpha)\tau_1V_c}\left[\frac{1}{\alpha}(1-e^{-\alpha t})-\frac{1}{\beta}\left(1-e^{-\beta t}\right)\right]$			(4.49)
		$C_P(t)=\frac{1}{\beta-\alpha}(C(\tau_1)[\beta e^{-\beta(t-\tau_1)}-\alpha e^{-\alpha(t-\tau_1)}]$ $+(k_{21}[C(\tau_1)+C_P(\tau_1)])\left(e^{-\alpha(t-\tau_1)}-e^{-\beta(t-\tau_1)}\right))$ $+\frac{F_2D}{(\beta-\alpha)\tau_2V_c}\left(\frac{k_{21}-\alpha}{\alpha}\left(1-e^{-\alpha(t-\tau_1)}\right)-\frac{k_{21}-\beta}{\beta}\left(1-e^{-\beta(t-\tau_1)}\right)\right)$	τ_1	$\tau_1+\tau_2$	(4.50)
		$C_P(t)=\frac{k_{12}}{\beta-\alpha}\left(\left(C(\tau_1)+\frac{k_{21}C_P(\tau_1)}{k_{21}-\alpha}\right)e^{-\alpha(t-\tau_1)}-(C(\tau_1)\right.$ $\left.+\frac{k_{21}C_P(\tau_1)}{k_{21}-\beta})e^{-\beta(t-\tau_1)}\right)+\frac{k_{12}F_2D}{(\beta-\alpha)\tau_2V_c}\left[\frac{1}{\alpha}\left(1-e^{-\alpha(t-\tau_1)}\right)-\frac{1}{\beta}\left(1-e^{-\beta(t-\tau_1)}\right)\right]$			(4.51)

		$C(t)=\frac{1}{\beta-\alpha}\left(C(\tau_1+\tau_2)\left[\beta e^{-\beta(t-\tau_1-\tau_2)}-\alpha e^{-\alpha(t-\tau_1-\tau_2)}\right]+(k_{21}[C(\tau_1+\tau_2)+C_{\mathrm{P}}(\tau_1+\tau_2)])\left(e^{-\alpha(t-\tau_1-\tau_2)}-e^{-\beta(t-\tau_1-\tau_2)}\right)\right)+\frac{F_3D}{(\beta-\alpha)\tau_3V_{\mathrm{c}}}\left(\frac{k_{21}-\alpha}{\alpha}\left(1-e^{-\alpha(t-\tau_1-\tau_2)}\right)-\frac{k_{21}-\beta}{\beta}\left(1-e^{-\beta(t-\tau_1-\tau_2)}\right)\right)$	$\tau_1+\tau_2$	$\tau_1+\tau_2+\tau_3$	(4.52)
		$C_{\mathrm{P}}(t)=\frac{k_{12}}{\beta-\alpha}\left(\left(C(\tau_1+\tau_2)+\frac{k_{21}C_{\mathrm{P}}(\tau_1+\tau_2)}{k_{21}-\alpha}\right)e^{-\alpha(t-\tau_1-\tau_2)}-(C(\tau_1+\tau_2)+\frac{k_{21}C_{\mathrm{P}}(\tau_1+\tau_2)}{k_{21}-\beta})e^{-\beta(t-\tau_1-\tau_2)}\right)+\frac{k_{12}F_3D}{(\beta-\alpha)\tau_3V_c}\left[\frac{1}{\alpha}\left(1-e^{-\alpha(t-\tau_1-\tau_2)}\right)-\frac{1}{\beta}\left(1-e^{-\beta(t-\tau_1-\tau_2)}\right)\right]$			(4.53)
		$C(t)=\frac{[C(\tau_1+\tau_2+\tau_3)(k_{21}-\alpha)+C_{\mathrm{P}}(\tau_1+\tau_2+\tau_3)k_{21}]e^{-\alpha(t-\tau_1-\tau_2-\tau_3)}}{\beta-\alpha}+\frac{[C(\tau_1+\tau_2+\tau_3)(k_{21}-\beta)+C_{\mathrm{P}}(\tau_1+\tau_2+\tau_3)k_{21}]e^{-\beta(t-\tau_1-\tau_2-\tau_3)}}{\alpha-\beta}$	$\tau_1+\tau_2+\tau_3$	∞	(4.54)
		$C_{\mathrm{P}}(t)=\frac{k_{12}}{\beta-\alpha}\left(\left(C(\tau_1+\tau_2+\tau_3)+\frac{k_{21}C_{\mathrm{P}}(\tau_1+\tau_2+\tau_3)}{k_{21}-\alpha}\right)e^{-\alpha(t-\tau_1-\tau_2-\tau_3)}-\left(C(\tau_1+\tau_2+\tau_3)+\frac{k_{21}C_{\mathrm{P}}(\tau_1+\tau_2+\tau_3)}{k_{21}-\beta}\right)e^{-\beta(t-\tau_1-\tau_2-\tau_3)}\right)$			(4.55)
		† In all cases, $\alpha+\beta=k_{12}+k_{21}+k_{10}$ and $\alpha\beta=k_{21}k_{10}$			(4.56)

Table 4.4 Solutions to the one-compartment model differential equations for 1, 2, or 3 successive input stages with explicit expressions

p	m	$C(t)$	$t_1 < t \leq t_2$		Eq.
1	1	$\frac{FD}{\tau V_d k_{el}}\left(1 - e^{-k_{el}t}\right)$	0	τ	(4.57)
		$\frac{FD}{\tau V_d k_{el}}\left(1 - e^{-k_{el}\tau}\right)e^{-k_{el}(t-\tau)}$	τ	∞	(4.58)
2	1	$\frac{F_1 D}{\tau_1 V_d k_{el}}\left(1 - e^{-k_{el}t}\right)$	0	τ_1	(4.59)
		$\frac{F_1 D}{\tau_1 V_d k_{el}}\left(1 - e^{-k_{el}\tau_1}\right)e^{-k_{el}(t-\tau_1)} + \frac{F_2 D}{\tau_2 V_d k_{el}}\left(1 - e^{-k_{el}(t-\tau_1)}\right)$	τ_1	$\tau_{1+}\tau_2$	(4.60)
		$\left[\frac{F_1 D}{\tau_1 V_d k_{el}}\left(1 - e^{-k_{el}\tau_1}\right)e^{-k_{el}\tau_2} + \frac{F_2 D}{\tau_2 V_d k_{el}}\left(1 - e^{-k_{el}\tau_2}\right)\right]e^{-k_{el}(t-\tau_1-\tau_2)}$	$\tau_{1+}\tau_2$	∞	(4.61)
3	1	$\frac{F_1 D}{\tau_1 V_d k_{el}}\left(1 - e^{-k_{el}t}\right)$	0	τ_1	(4.62)
		$\frac{F_1 D}{\tau_1 V_d k_{el}}\left(1 - e^{-k_{el}\tau_1}\right)e^{-k_{el}(t-\tau_1)} + \frac{F_2 D}{\tau_2 V_d k_{el}}\left(1 - e^{-k_{el}(t-\tau_1)}\right)$	τ_1	$\tau_{1+}\tau_2$	(4.63)
		$\left[\frac{F_1 D}{\tau_1 V_d k_{el}}\left(1 - e^{-k_{el}\tau_1}\right)e^{-k_{el}\tau_2} + \frac{F_2 D}{\tau_2 V_d k_{el}}\left(1 - e^{-k_{el}\tau_2}\right)\right]e^{-k_{el}(t-\tau_1-\tau_2)} + \frac{F_3 D}{\tau_3 V_d k_{el}}\left(1 - e^{-k_{el}(t-\tau_1-\tau_2)}\right)$	$\tau_{1+}\tau_2$	$\tau_{1+}\tau_{2+}\tau_3$	(4.64)
		$\left(\left[\frac{F_1 D}{\tau_1 V_d k_{el}}\left(1 - e^{-k_{el}\tau_1}\right)e^{-k_{el}\tau_2} + \frac{F_2 D}{\tau_2 V_d k_{el}}\left(1 - e^{-k_{el}\tau_2}\right)\right]e^{-k_{el}\tau_3} + \frac{F_3 D}{\tau_3 V_d k_{el}}\left(1 - e^{-k_{el}\tau_3}\right)\right)e^{-k_{el}(t-\tau_1-\tau_2-\tau_3)}$	$\tau_{1+}\tau_{2+}\tau_3$	∞	(4.65)

Table 4.5 Solutions to the two-compartment model differential equations for 1, 2, or 3 successive input stages with explicit expressions

P	m	$C(t)$, $C_P(t)$ †	$t_1 < t \leq t_2$		Eq.
1	2	$C(t) = \frac{FD}{(\alpha-\beta)\tau V_c}\left[(k_{21}-\beta)\frac{1-e^{-\beta t}}{\beta} - (k_{21}-\alpha)\frac{1-e^{-at}}{a}\right]$	0	τ	(4.66)
		$C_P(t) = \frac{FDk_{12}}{(\alpha-\beta)\tau V_c}\left(\frac{1-e^{-\beta t}}{\beta} - \frac{1-e^{-\alpha t}}{\alpha}\right)$			(4.67)
		$C(t) = \frac{FD}{(\alpha-\beta)\tau V_c}\left[(k_{21}-\beta)\frac{1-e^{-\beta\tau}}{\beta}e^{-\beta(t-\tau)} - (k_{21}-\alpha)\frac{1-e^{-a\tau}}{a}e^{-\alpha(t-\tau)}\right]$	τ	∞	(4.68)
		$C_P(t) = \frac{FD}{\tau V_c}\left[\frac{1-e^{-\beta\tau}}{\beta}e^{-\beta(t-\tau)} - \frac{1-e^{-a\tau}}{a}e^{-\alpha(t-\tau)}\right]$			(4.69)
2	2	$C(t) = \frac{F_1D}{(\alpha-\beta)\tau_1 V_c}\left[(k_{21}-\beta)\frac{1-e^{-\beta t}}{\beta} - (k_{21}-\alpha)\frac{1-e^{-at}}{a}\right]$	0	τ_1	(4.70)
		$C_P(t) = \frac{F_1Dk_{12}}{(\alpha-\beta)\tau_1 V_c}\left(\frac{1-e^{-\beta t}}{\beta} - \frac{1-e^{-\alpha t}}{\alpha}\right)$			(4.71)
		$C(t) = \frac{F_1D}{(\alpha-\beta)\tau_1 V_c}\left[(k_{21}-\beta)\frac{1-e^{-\beta\tau_1}}{\beta}e^{-\beta(t-\tau_1)} - (k_{21}-\alpha)\frac{1-e^{-a\tau_1}}{a}e^{-\alpha(t-\tau_1)}\right]$ $+\frac{F_2D}{(\alpha-\beta)\tau_2 V_c}\left[(k_{21}-\beta)\frac{1-e^{-\beta(t-\tau_1)}}{\beta} - (k_{21}-\alpha)\frac{1-e^{-\alpha(t-\tau_1)}}{\alpha}\right]$	τ_1	$\tau_1 + \tau_2$	(4.72)
		$C_P(t) = \frac{F_1Dk_{12}}{(\alpha-\beta)\tau_1 V_c}\left[\frac{1-e^{-\beta\tau_1}}{\beta}e^{-\beta(t-\tau_1)} - \frac{1-e^{-a\tau_1}}{a}e^{-\alpha(t-\tau_1)}\right] + \frac{F_2Dk_{12}}{(\alpha-\beta)\tau_2 V_c}\left[\frac{1-e^{-\beta(t-\tau_1)}}{\beta} - \frac{1-e^{-\alpha(t-\tau_1)}}{\alpha}\right]$			(4.73)
		$C(t) = \frac{D}{(\alpha-\beta)V_c}\left[(k_{21}-\beta)\left(\frac{F_1}{\tau_1}\frac{1-e^{-\beta\tau_1}}{\beta}e^{-\beta\tau_2} + \frac{F_2}{\tau_2}\frac{1-e^{-\beta\tau_2}}{\beta}\right)e^{-\beta(t-\tau_1-\tau_2)}\right.$ $\left.-(k_{21}-\alpha)\left(\frac{F_1}{\tau_1}\frac{1-e^{-\alpha\tau_1}}{\alpha}e^{-\alpha\tau_2} + \frac{F_2}{\tau_2}\frac{1-e^{-\alpha\tau_2}}{\alpha}\right)e^{-\alpha(t-\tau_1-\tau_2)}\right]$	$\tau_1 + \tau_2$	∞	(4.74)
		$C_P(t) = \frac{Dk_{12}}{(\alpha-\beta)V_c}\left[\left(\frac{F_1}{\tau_1}\frac{1-e^{-\beta\tau_1}}{\beta}e^{-\beta\tau_2} + \frac{F_2}{\tau_2}\frac{1-e^{-\beta\tau_2}}{\beta}\right)e^{-\beta(t-\tau_1-\tau_2)}\right.$ $\left.-\left(\frac{F_1}{\tau_1}\frac{1-e^{-\alpha\tau_1}}{\alpha}e^{-\alpha\tau_2} + \frac{F_2}{\tau_2}\frac{1-e^{-\alpha\tau_2}}{\alpha}\right)e^{-\alpha(t-\tau_1-\tau_2)}\right]$			(4.75)

(continued)

Table 4.5 (continued)

P	m	$C(t)$, $C_P(t)$ †	$t_1 < t \leq t_2$		Eq.
3	2	$C(t)=\frac{F_1D}{(\alpha-\beta)\tau_1V_c}\left[\frac{(k_{21}-\beta)\left(1-e^{-\beta t}\right)}{\beta}-\frac{(k_{21}-\alpha)(1-e^{-\alpha t})}{\alpha}\right]$	0	τ_1	(4.76)
		$C_P(t)=\frac{F_1Dk_{12}}{(\alpha-\beta)\tau_1V_c}\left(\frac{1-e^{-\beta t}}{\beta}-\frac{1-e^{-\alpha t}}{\alpha}\right)$			(4.77)
		$C(t)=\frac{F_1D}{(\alpha-\beta)\tau_1V_c}\left[(k_{21}-\beta)\frac{1-e^{-\beta\tau_1}}{\beta}e^{-\beta(t-\tau_1)}-(k_{21}-\alpha)\frac{1-e^{-\alpha\tau_1}}{a}e^{-\alpha(t-\tau_1)}\right]$ $+\frac{F_2D}{(\alpha-\beta)\tau_2V_c}\left[(k_{21}-\beta)\frac{1-e^{-\beta(t-\tau_1)}}{\beta}-(k_{21}-\alpha)\frac{1-e^{-\alpha(t-\tau_1)}}{\alpha}\right]$	τ_1	$\tau_1+\tau_2$	(4.78)
		$C_P(t)=\frac{F_1Dk_{12}}{(\alpha-\beta)\tau_1V_c}\left[\frac{1-e^{-\beta\tau_1}}{\beta}e^{-\beta(t-\tau_1)}-\frac{1-e^{-\alpha\tau_1}}{a}e^{-\alpha(t-\tau_1)}\right]+\frac{F_2Dk_{12}}{(\alpha-\beta)\tau_2V_c}\left[\frac{1-e^{-\beta(t-\tau_1)}}{\beta}-\frac{1-e^{-\alpha(t-\tau_1)}}{\alpha}\right]$			(4.79)
		$C(t)=\frac{1}{\alpha-\beta}\left([C(\tau_1+\tau_2)(k_{21}-\beta)+C_P(\tau_1+\tau_2)k_{21}]e^{-\beta(t-\tau_1-\tau_2)}-[C(\tau_1+\tau_2)(k_{21}-\alpha)\right.$ $\left.+C_P(\tau_1+\tau_2)k_{21}]e^{-\alpha(t-\tau_1-\tau_2)}\right)+\frac{F_3D}{(\alpha-\beta)\tau_3V_c}$ $\times\left(\frac{k_{21}-\beta}{\beta}\left(1-e^{-\beta(t-\tau_1-\tau_2)}\right)-\frac{k_{21}-\alpha}{\alpha}\left(1-e^{-\alpha(t-\tau_1-\tau_2)}\right)\right)$	$\tau_1+\tau_2$	$\tau_1+\tau_2+\tau_3$	(4.80)
		$C_P(t)=\frac{k_{12}}{\alpha-\beta}\left[\frac{C(\tau_1+\tau_2)(k_{21}-\beta)+k_{21}C_P(\tau_1+\tau_2)}{k_{21}-\beta}e^{-\beta(t-\tau_1-\tau_2)}\right.$ $\left.-\frac{C(\tau_1+\tau_2)(k_{21}-\alpha)+k_{21}C_P(\tau_1+\tau_2)}{k_{21}-\alpha}e^{-a(t-\tau_1-\tau_2)}\right]+\frac{k_{12}F_3D}{(\alpha-\beta)\tau_3V_c}$ $\times\left[\frac{1-e^{-\beta(t-\tau_1-\tau_2)}}{\beta}-\frac{1-e^{-a(t-\tau_1-\tau_2)}}{\alpha}\right]$			(4.81)

		$C(t) = \frac{1}{\alpha-\beta}\left([C(\tau_1+\tau_2+\tau_3)(k_{21}-\beta) + C_P(\tau_1+\tau_2+\tau_3)k_{21}]e^{-\beta(t-\tau_1-\tau_2-\tau_3)} - [C(\tau_1+\tau_2+\tau_3)(k_{21}-\alpha) + C_P(\tau_1+\tau_2+\tau_3)k_{21}]e^{-\alpha(t-\tau_1-\tau_2-\tau_3)}\right)$	$\tau_1+\tau_2+\tau_3$	∞	(4.82)
		$C_P(t) = \frac{k_{12}}{\beta-\alpha}\left(\left(C(\tau_1+\tau_2+\tau_3) + \frac{k_{21}C_P(\tau_1+\tau_2+\tau_3)}{k_{21}-\alpha}\right)e^{-\alpha(t-\tau_1-\tau_2-\tau_3)} - \left(C(\tau_1+\tau_2+\tau_3) + \frac{k_{21}C_P(\tau_1+\tau_2+\tau_3)}{k_{21}-\beta}\right)e^{-\beta(t-\tau_1-\tau_2-\tau_3)}\right)$			(4.83)
		† In all cases, $\alpha + \beta = k_{12} + k_{21} + k_{10}$ and $\alpha\beta = k_{21}k_{10}$			(4.84)

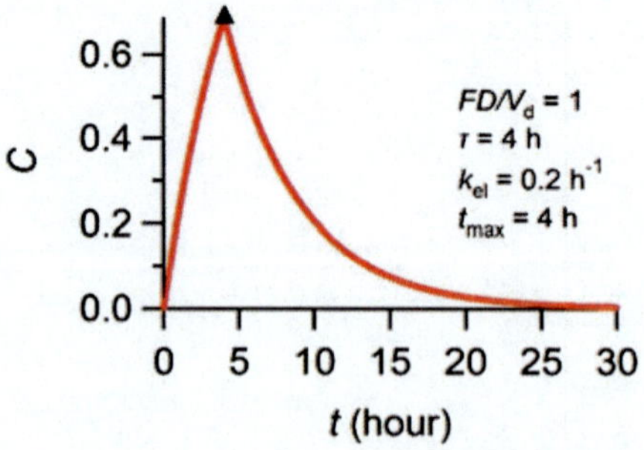

Fig. 4.3 Simulated drug concentration in the blood for an orally administered formulation that follows a zero-order absorption process for a finite time in a one-compartment model, see Eqs. (4.57)–(4.58). Maximum concentration is reached at the end of the absorption stage, i.e., at time τ. Model parameter values can be seen next to the curve

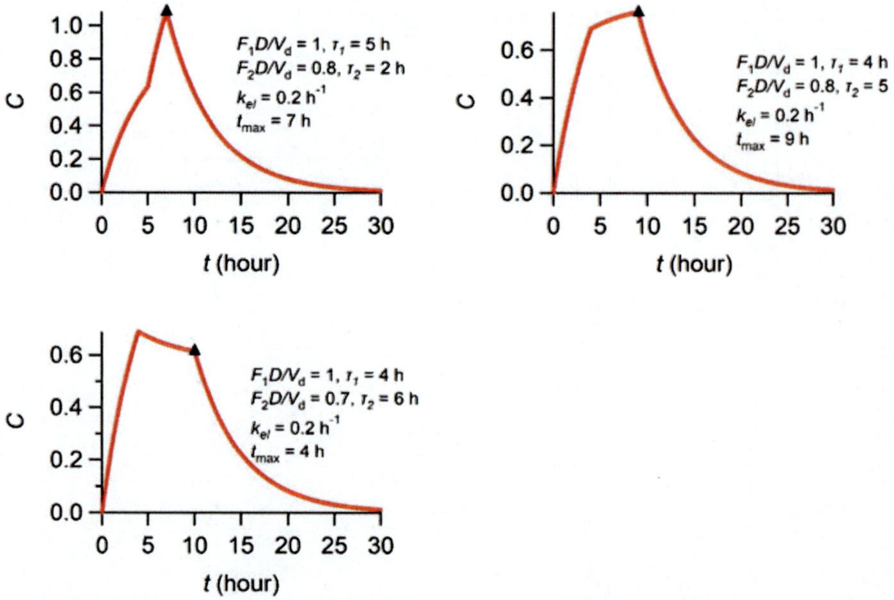

Fig. 4.4 Simulated drug concentration in the blood for an orally administered formulation that follows a zero-order absorption process for a finite time in a one-compartment model, see Eqs. (4.59)–(4.61). Maximum concentration is reached either at the end of the first absorption stage or after both stages. Model parameter values can be seen next to each curve

The simulations presented in Figs. 4.3, 4.4, 4.5, 4.6, 4.7, and 4.8 resemble real-life data. This is in contrast to the use of the absorption rate constant, which cannot capture the dynamics of the complex absorption processes. The physiologically sound consecutive passive absorption steps under the sink conditions premise, lead to C, t profiles which are actually observed in pharmacokinetic studies.

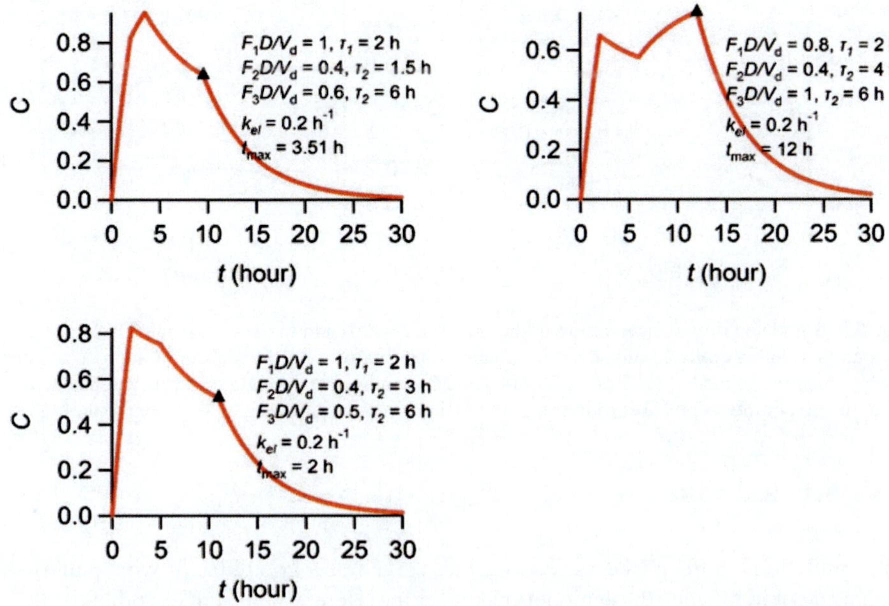

Fig. 4.5 Simulated drug concentration in the blood for an orally administered formulation that follows a zero-order absorption process for a finite time in a one-compartment model, see Eqs. (4.62)–(4.65). Maximum concentration is reached either at the end of the first, the second, or the third absorption stages. Model parameter values can be seen next to each curve. This model is flexible enough to allow two maxima in the drug concentration, one global and one local

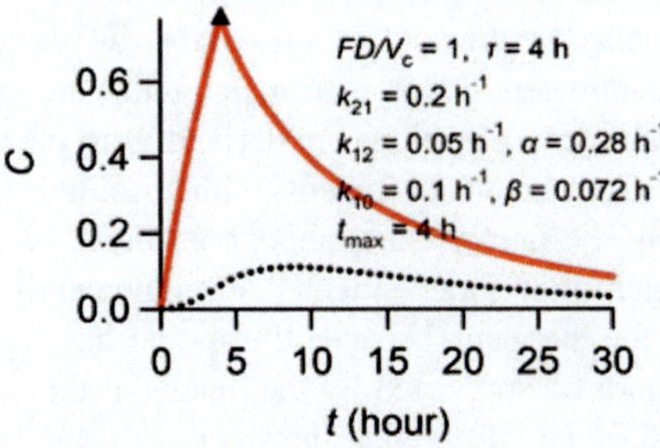

Fig. 4.6 Simulated drug concentration in the blood (red solid) and in the peripheral (black dotted) compartment for an orally administered formulation that follows a zero-order absorption process for a finite time in a two-compartment model, see Eqs. (4.66)–(4.69). Maximum concentration is reached at the end of the absorption stage, i.e., at time τ. Model parameter values can be seen next to the curve

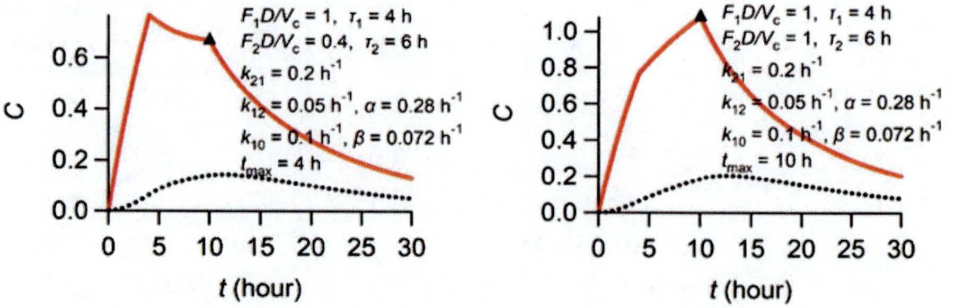

Fig. 4.7 Simulated drug concentration in the blood (red solid) and in the peripheral (black dotted) compartment for an orally administered formulation that follows zero-order absorption processes in a two-compartment model, see Eqs. (4.70)–(4.75). Maximum concentration is reached at the end of either the first or the second absorption stage. Model parameter values can be seen next to the curve

4.3 Fits to Data

The simulations in the previous sections have given us a good idea of what patterns in pharmacokinetic data these models can describe. Here, we present examples of fits on real data. Since this section is very well suited for teaching purposes, we follow a didactic style in the analysis of the data.

We will try a variety of models on the same sets in an attempt to identify the best choice. The quality of a fit for a certain model on a given set of data can be established by comparing the sum of squares of deviations, χ^2, between calculated values from the models and experimental points; a lower value for χ^2 signifies a closer match between the data and the model. So, the model with a lowest χ^2 is usually the more appropriate one for a given drug formulation, as long as the parameter values determined are physical. Other criteria of model quality are the uncertainties of the calculated parameters as well as correlations among the parameters. Parameter uncertainties are calculated directly from parameter variances. Parameter correlations are derived from parameter covariances and corresponding variances. Correlation coefficients with absolute values close to 1 indicate strong correlations among the pair of parameters considered, which implies that these parameters cannot be determined independently or, equivalently, that the model is not appropriate. Most often this happens because the model has far too many parameters whose determination cannot be supported by the details in the data. Each fit is accompanied by a plot of residuals, i.e., differences between data and calculated values. These plots are helpful in identifying pattern differences between data and models.

N.B. Pharmacokinetic data are frequently sparse because of the cost and inconvenience incurred by the need to collect blood samples and analyze them. Furthermore, great variability is seen among individual volunteers. On top of all these difficulties and uncertainties, analysis of published data often is made even more complicated and inaccurate because of the digitization step of printed curves.

First, we focus on the elimination phase data as a preliminary check for identifying the best disposition model and determining the elimination rate constant, k_{el} or

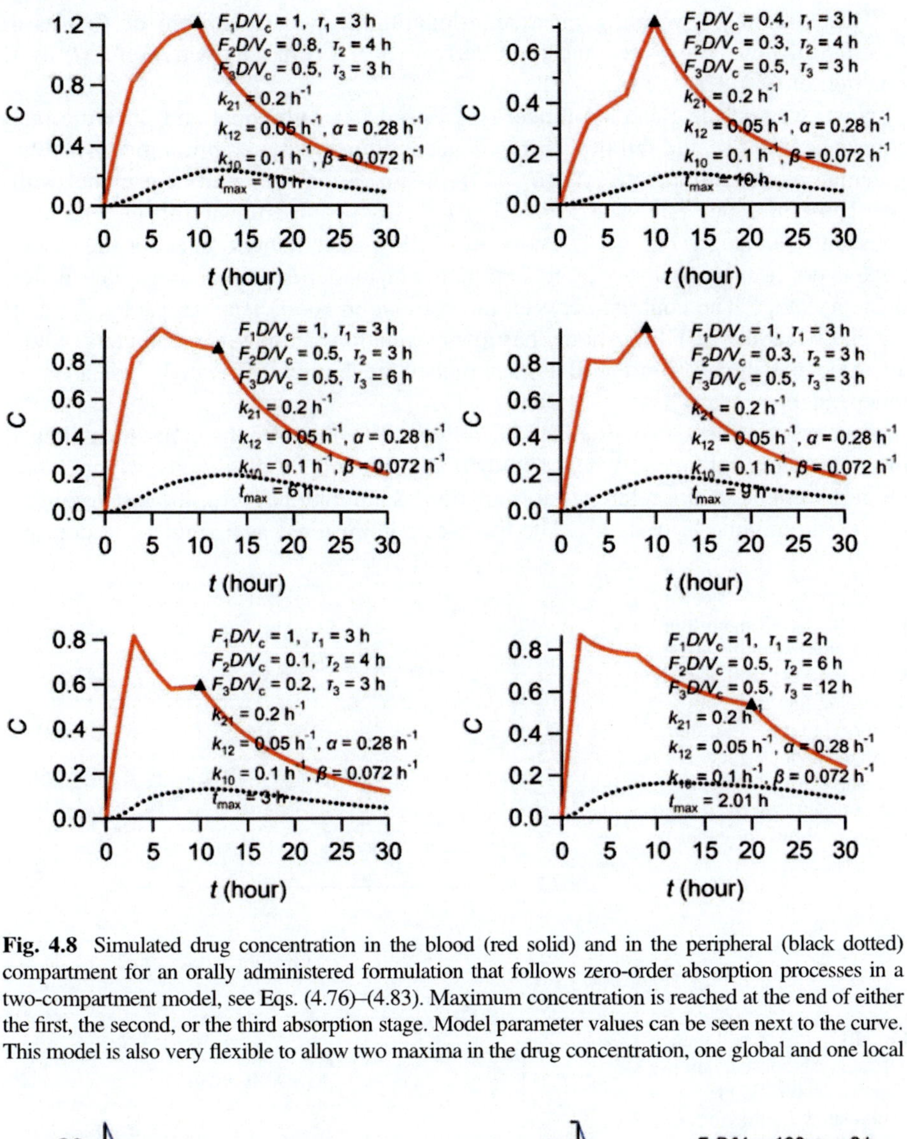

Fig. 4.8 Simulated drug concentration in the blood (red solid) and in the peripheral (black dotted) compartment for an orally administered formulation that follows zero-order absorption processes in a two-compartment model, see Eqs. (4.76)–(4.83). Maximum concentration is reached at the end of either the first, the second, or the third absorption stage. Model parameter values can be seen next to the curve. This model is also very flexible to allow two maxima in the drug concentration, one global and one local

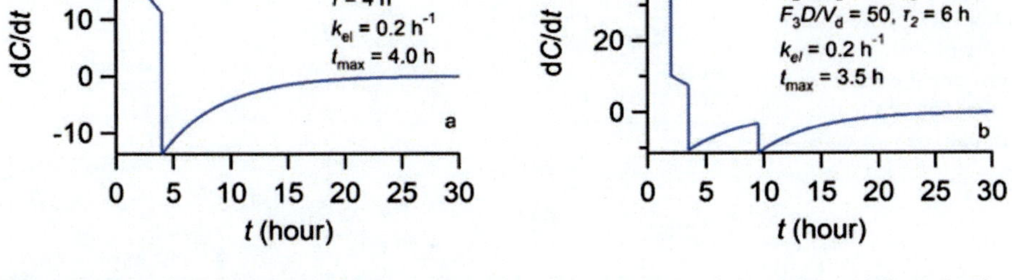

Fig. 4.9 Plots of the derivative dC/dt as a function of time. (**a**) $t_{\max} = \tau = 4.0$ h and the derivative was calculated from Eqs. (4.57) and (4.58). (**b**) $t_{\max} = 3.5$ h $< \tau = 9.5$ h and the derivative was calculated from Eqs. (4.62)–(4.65)

β. This is done by working on a semi-logarithmic plot (logarithm of $C(t)$ as a function of time, Fig. 4.10) and on a derivative plot (time derivative of $C(t)$ as a function of $C(t)$) [2], see Fig. 4.11.

Next, we test the Bateman model (see Fig. 4.12) with equal and differing rate constants. The rise and fall of the drug concentration are more abrupt than what the Bateman model can achieve. Then, we try a simple, one-compartment model with one constant input rate (see Fig. 4.13) or a two-compartment model with one constant input stage (Fig. 4.14). Based on the low uncertainty ranges for the model parameters it appears that the one-compartment model is performing much better than any other. The contrast between the correlation coefficients in Tables 4.6 and 4.7 show that the PBFTPK model has low correlations among its parameters, while all three parameters for the Bateman model are highly correlated, i.e., are not independent.

An analysis of cyclosporine data [4] follows. We analyzed the experimental data of the fundamental bioequivalence study under fast and fed conditions, which led to the replacement of the reference formulation (Sandimmune) with the test formulation (Sandimmune Neoral) [4]. The test was administered as a single oral dose of

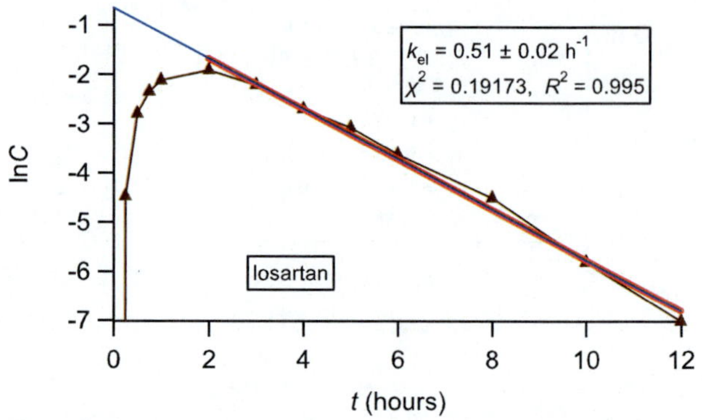

Fig. 4.10 Semi-logarithmic plot of pharmacokinetic data for losartan [3]. A linear regression line is drawn from t_{max} to the end of the available data. A good estimate of the elimination rate constant is derived from the fit, as shown in the inset

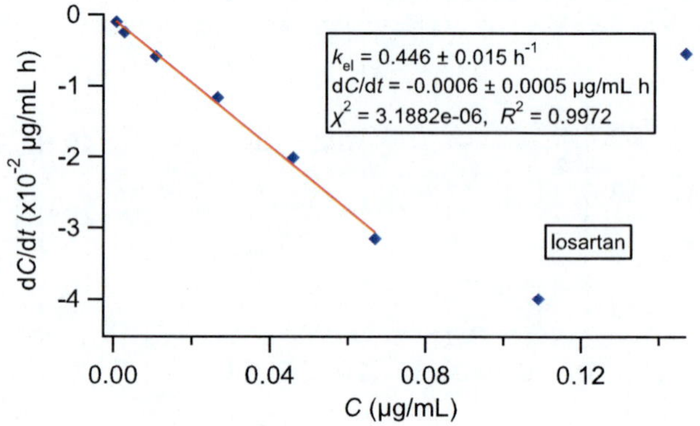

Fig. 4.11 Pharmacokinetic data analysis for losartan [3] to establish the simple exponential elimination phase rate constant. Moreover, the intercept ± SD estimate overlaps zero, a finding verifying the reliable estimation of the terminal elimination rate constant

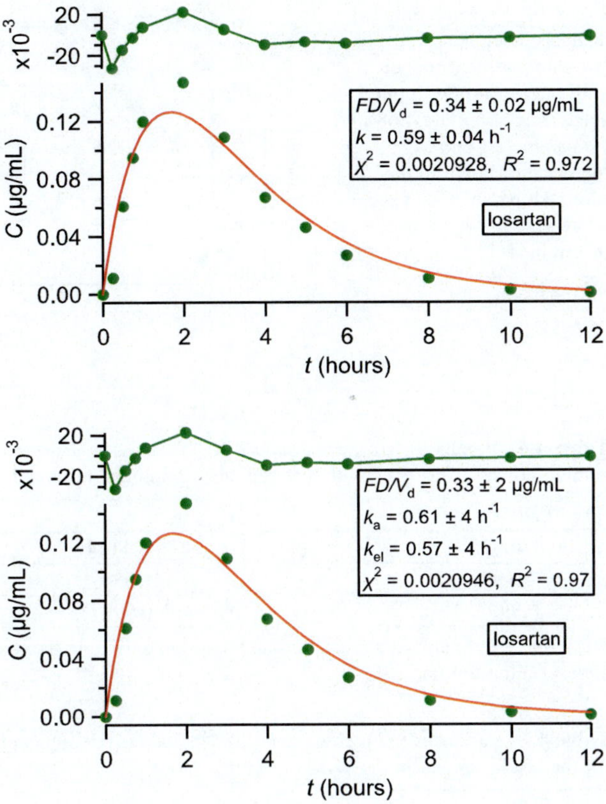

Fig. 4.12 Losartan data [3] analysis based on the Bateman equation, with equal (Eq. 1.3) or unequal (Eq. 1.2) absorption and elimination rate constants

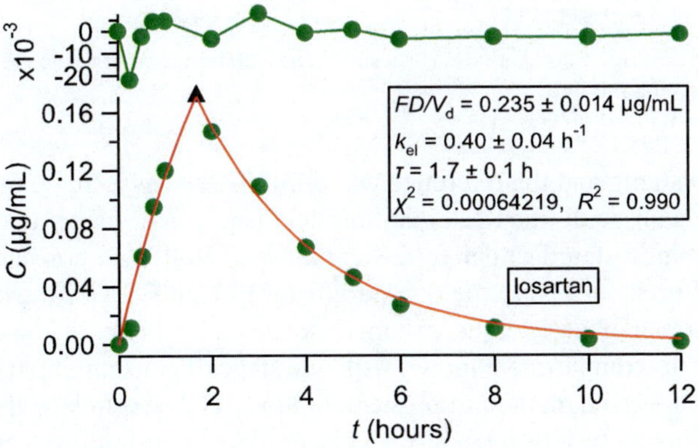

Fig. 4.13 Losartan data [3] analysis using a one-compartment model with one zero-order input stage (Eqs. 4.57 and 4.58). Fit parameters and fit metrics are shown in the inset

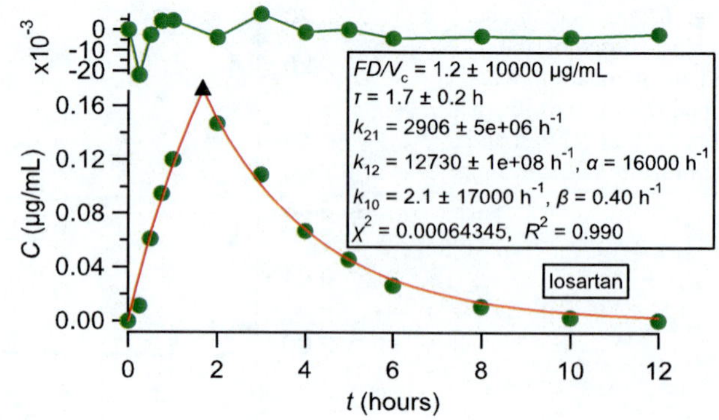

Fig. 4.14 Losartan data [3] analysis using a two-compartment model with one zero-order input stage (Eqs. 4.66 and 4.68). Fit parameters and fit metrics are shown in the inset. All parameters, except τ, are very poorly determined

Table 4.6 Variances (σ_i^2), covariances (σ_{ij}^2), and correlations (r_{ij}^2) among parameters for the results shown in Fig. 4.12 with different absorption and elimination rate constants

$i,j = FD/V_d$	k_a	k_{el}
$\sigma_1^2 = 4.40422$	$\sigma_{12}^2 = -8.1144$	$\sigma_{13}^2 = 7.65623$
$r_{12}^2 = -0.999882$	$\sigma_2^2 = 14.9536$	$\sigma_{23}^2 = -14.1056$
$r_{13}^2 = 0.999924$	$r_{23}^2 = -0.99978$	$\sigma_3^2 = 13.3115$

Subscripts *i* and *j* are used as a shorthand notation to refer to the model parameters in the order they appear on the top line

Table 4.7 Variances (σ_i^2), covariances (σ_{ij}^2), and correlation coefficients (r_{ij}^2) among parameters for the results shown in Fig. 4.13

$i,j = FD/V_d$	k_{el}	τ
$\sigma_1^2 = 0.000204293$	$\sigma_{12}^2 = 0.000439226$	$\sigma_{13}^2 = 0.000920279$
$r_{12}^2 = 0.852454$	$\sigma_2^2 = 0.00129951$	$\sigma_{23}^2 = 0.00229837$
$r_{13}^2 = 0.643437$	$r_{23}^2 = 0.63715$	$\sigma_3^2 = 0.0100132$

Subscripts *i* and *j* are used as a shorthand notation to refer to the model parameters in the order given on the top line

180 mg and the reference was administered as a single oral dose of 300 mg. We start again with the Bateman model, Eq. (1.2), using data for the test formulation administered under fasted conditions. Both rate constants are nearly identical, see Fig. 4.15. Using the one-parameter model, Eq. (1.3), we get the same χ^2, but much lower parameter uncertainty, so this is a better fit, see Fig. 4.16. We then try a one-compartment model with one stage of constant input for finite time τ. Figure 4.17 shows the marked improvement in χ^2. The next fit is with a two-compartment model, Eqs. (4.66)–(4.69), shows a discernible deterioration over the previous fit (Fig. 4.18).

Next, we work on another set of cyclosporine data [4], namely, the reference formulation under fed conditions, which has a distinctly different pattern. First, we try the Bateman equation with unequal and then equal rate constants. The rather poor results are shown in Fig. 4.19.

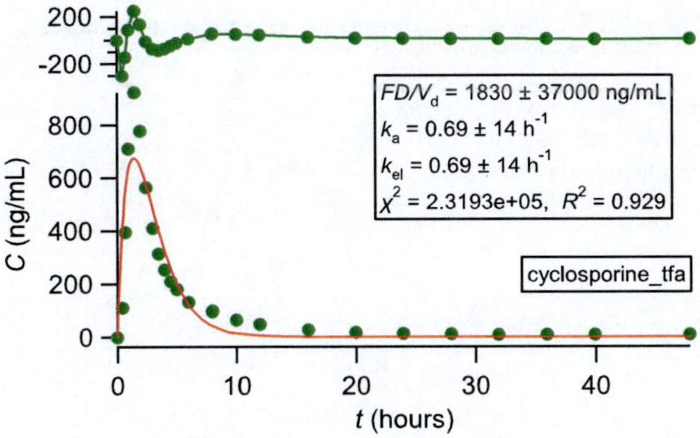

Fig. 4.15 Bateman fit to cyclosporine test formulation data under fasted conditions [4]

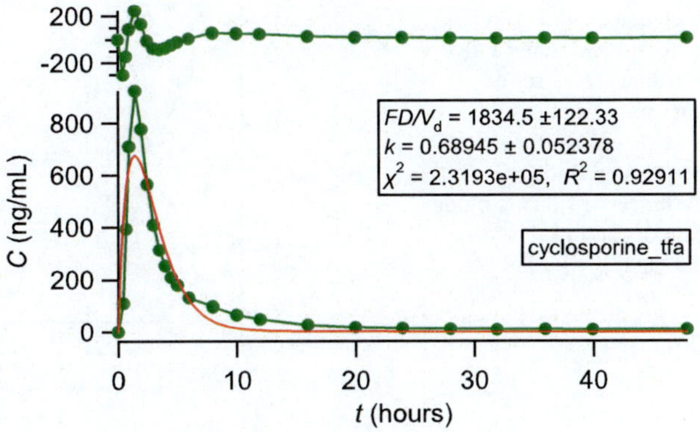

Fig. 4.16 Simplified (one-rate-parameter) Bateman fit (Eq. 1.3) to cyclosporine test formulation data under fasted conditions [4]

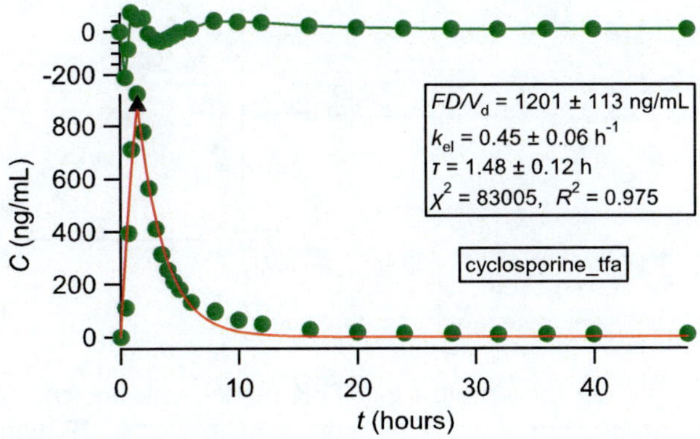

Fig. 4.17 One compartment, one constant input rate for time τ fit (Eqs. 4.57–4.58) to cyclosporine test formulation data under fasted conditions [4]

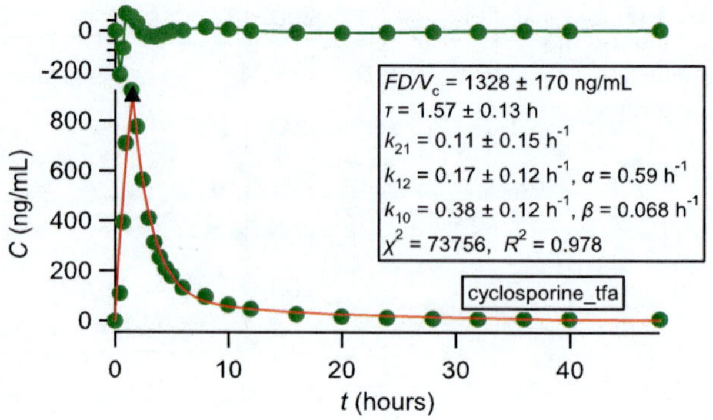

Fig. 4.18 Two-compartment, one constant input rate for time τ fit (Eqs. 4.66–4.69) to cyclosporine test formulation data under fasted conditions [4]

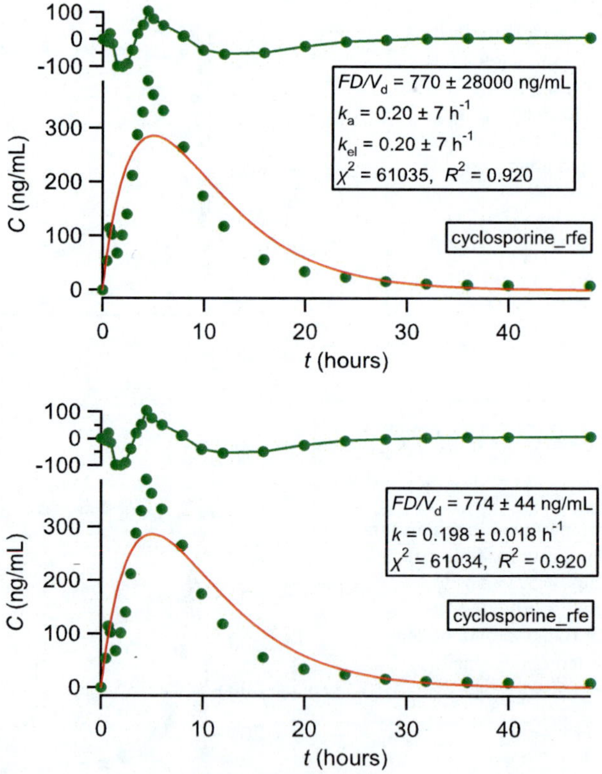

Fig. 4.19 Fit with the Bateman model with different and equal rate constants for a set of pharmacokinetic data of a reference formulation administered to fed volunteers [4]. This model cannot capture the form of the data

We test the simplest PBFTPK model with one constant input and one compartment, see Fig. 4.20. The improvement over the Bateman model is clear, but the pattern is not captured by the functional form of $C(t)$ given by the Eqs. (4.57)–(4.58). Next, we implement additional input stages. Figure 4.21 shows that the addition of a second input stage improves the fit. Figure 4.22 shows a model fit with three

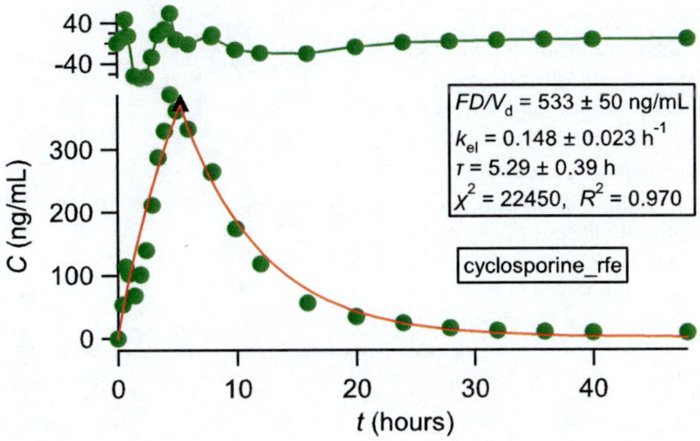

Fig. 4.20 Fit with one input stage to a one-compartment model disposition to cyclosporine reference formulation data under fed conditions [4]

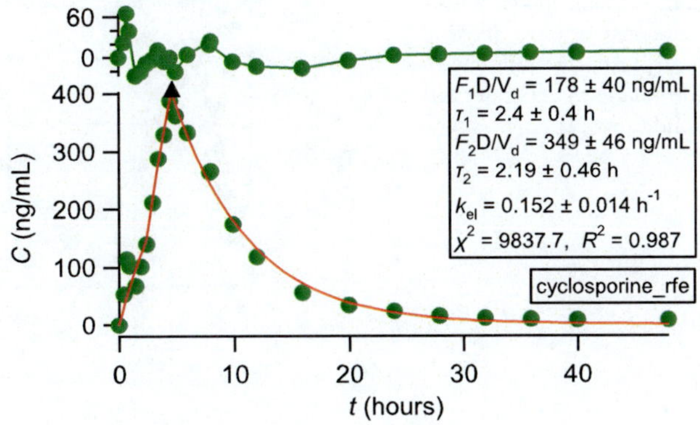

Fig. 4.21 One-compartment model with two successive constant input stages fit to cyclosporine reference formulation data under fed conditions [4]

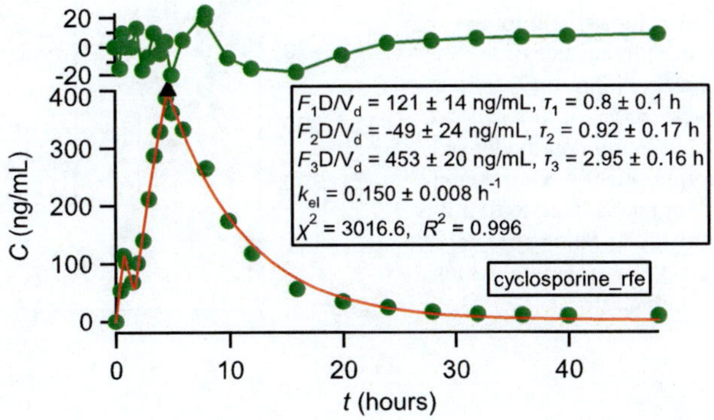

Fig. 4.22 One-compartment model with three successive constant input stages fit to cyclosporine reference formulation data under fed conditions [4]

successive constant input stages. At first glance this is the best fit so far. Details of the data are captured by the model, but something is not right. One input rate is negative, which cannot be the case in a real system. By imposing a constraint on the input rates requiring them to be positive, we get the fit shown in Fig. 4.23, which indicates that during the second stage absorption has been discontinued.

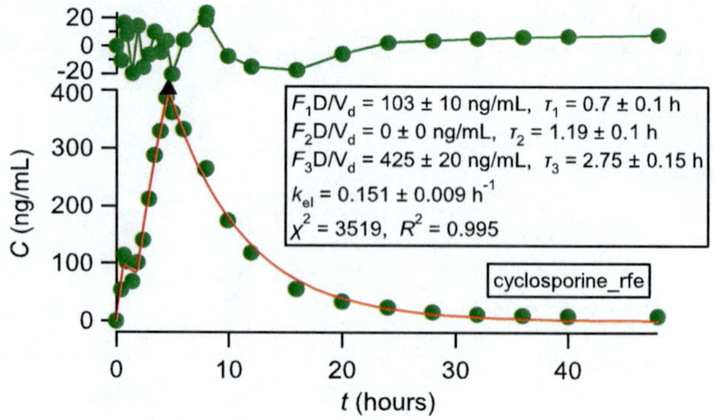

Fig. 4.23 One-compartment model with three constrained successive constant input stages fit to cyclosporine reference formulation data under fed conditions [4]; the second input rate is set at 0 to avoid an unphysical negative value (see Fig. 4.22)

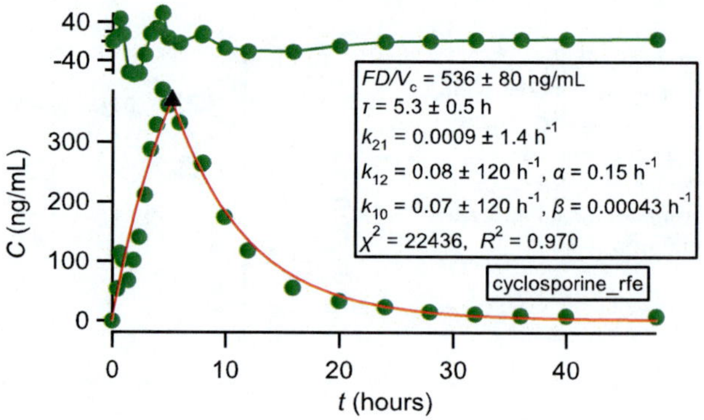

Fig. 4.24 Two-compartment model with one constant input stage fit to cyclosporine reference formulation data under fed conditions [4]

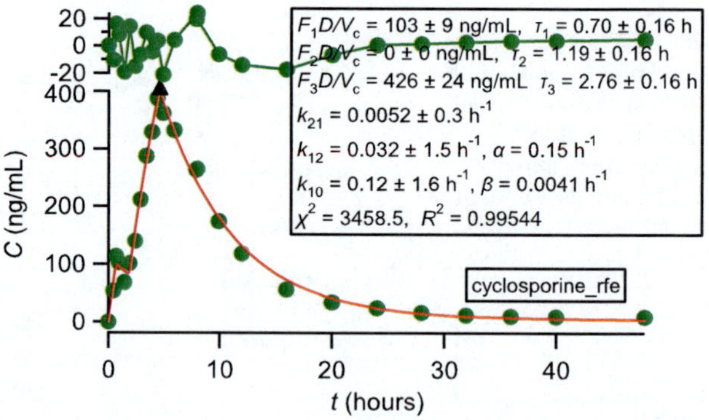

Fig. 4.25 Fit with a two-compartment model using three successive constant input rates to cyclosporine reference formulation data under fed conditions [4]. The second input rate is fixed to 0; if it is allowed to adjust, the value selected is negative, which is unnatural

Next, we attempt to fit the same data with two-compartment models. The result for one input stage is shown in Fig. 4.24. The fit is not very good, but, more importantly, one rate constant, k_{10} has a negative value, which is certainly unphysical. Using three successive inputs in a two-compartment model gives an improved fit, see Fig. 4.25. To make the fit yield meaningful results we had to fix the second input rate to 0. Allowing this parameter to adjust resulted in a meaningless negative value. Even so, rate constants arrived at have very high uncertainties. Overall the best results were reached with the one-compartment model with three successive input rates, of which the second did not contribute to the drug absorption, Fig. 4.23.

Figure 4.25 shows the complex absorption of cyclosporine from the reference formulation under fed conditions; in fact, the best fit (Fig. 4.23) corresponds to a model with three successive fluctuating input rates of total duration of 4.6 h. These results are indicative of the erratic absorption of cyclosporine from the reference formulation in the presence of food. These findings are related to the hydrophobic nature of cyclosporine and the pharmaceutical differences of the two formulations, namely, the test formulation is a microemulsion (Fig. 4.17), while the reference formulation is a solution of cyclosporine in olive oil (Fig. 4.23).

Ibuprofen [5] pharmacokinetic data are analyzed first with the Bateman model with equal or unequal rate constants (Fig. 4.26). The fit is adequate, but we try the FAT concept. Figure 4.27 shows a fit with one constant input rate, but using two constant input rates gives a much better fit (Fig. 4.28). All metrics (χ^2, R^2, parameter uncertainties, correlations, Table 4.8) show a marked advantage over the previous

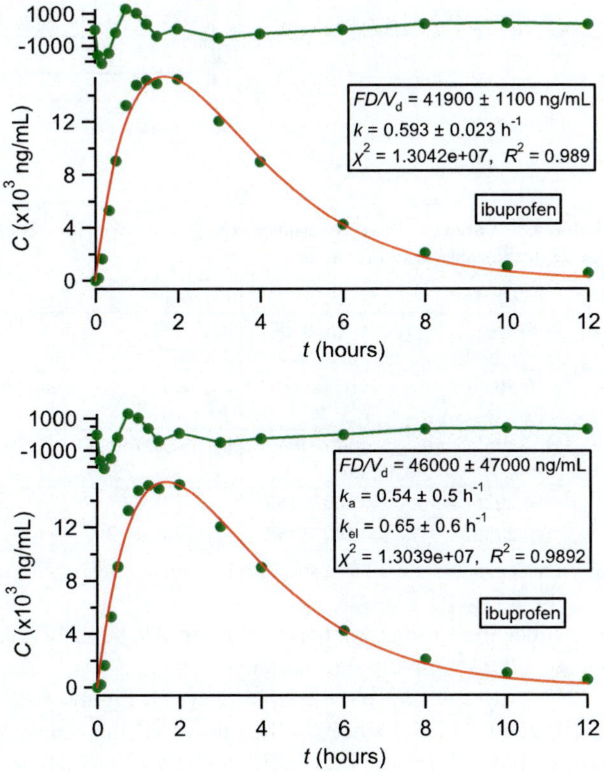

Fig. 4.26 The Bateman equation with equal or unequal rate constants fit ibuprofen data [5] only approximately

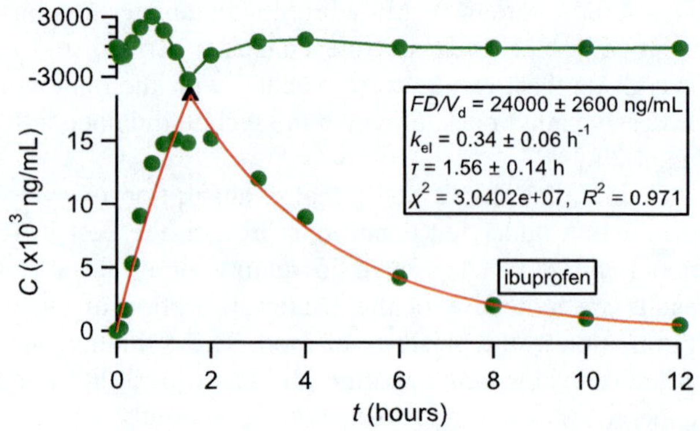

Fig. 4.27 One-compartment model with one constant input rate misses some of the features in the ibuprofen data [5]

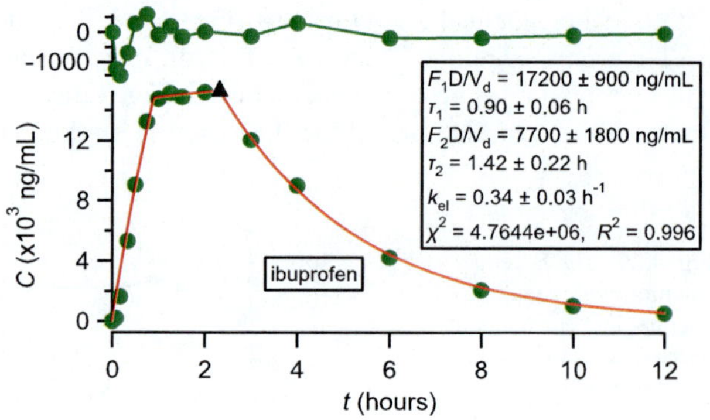

Fig. 4.28 One-compartment model with two successive constant rate inputs describes ibuprofen data [5] better than any other of the models tried

Table 4.8 Variances (σ_i^2), covariances (σ_{ij}^2), and correlation coefficients (r_{ij}^2) among parameters for the results shown in Fig. 4.28

$i,j = F_1D/V_d$	τ_1	F_2D/V_d	τ_2	k_{el}
$\sigma_1^2 = 818{,}165$	$\sigma_{12}^2 = 45.2624$	$\sigma_{13}^2 = -343{,}677$	$\sigma_{14}^2 = 58.4508$	$\sigma_{15}^2 = 9.45129$
$r_{12}^2 = 0.809764$	$\sigma_2^2 = 0.0038187$	$\sigma_{23}^2 = -39.3701$	$\sigma_{24}^2 = 0.000384$	$\sigma_{25}^2 = 0.000176$
$r_{13}^2 = -0.209964$	$r_{23}^2 = -0.352065$	$\sigma_3^2 = 3.274e+06$	$\sigma_{34}^2 = 224.181$	$\sigma_{35}^2 = 50.5863$
$r_{14}^2 = 0.292587$	$r_{24}^2 = 0.028153$	$r_{34}^2 = 0.560917$	$\sigma_4^2 = 0.0487786$	$\sigma_{45}^2 = 0.004541$
$r_{15}^2 = 0.299533$	$r_{25}^2 = 0.081855$	$r_{35}^2 = 0.801347$	$r_{45}^2 = 0.589435$	$\sigma_5^2 = 0.0012168$

Subscripts i and j are used as a shorthand notation to refer to the model parameters

fits. Since the elimination part of the curve is described very well, there is no point in trying a two-compartment model.

We now consider paracetamol data [5]. Figure 4.29 shows the shortcomings of the Bateman model since the fit misses all the points on the rising section of the curve. Fits with one (Fig. 4.30) and two constant input rates (Fig. 4.31) show

Fig. 4.29 Bateman model fit to paracetamol data [5]

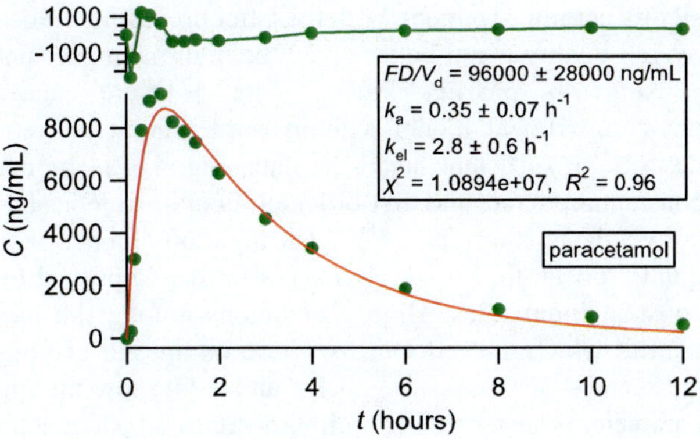

Fig. 4.30 Fit with one-compartment, one constant input rate model to paracetamol data [5]

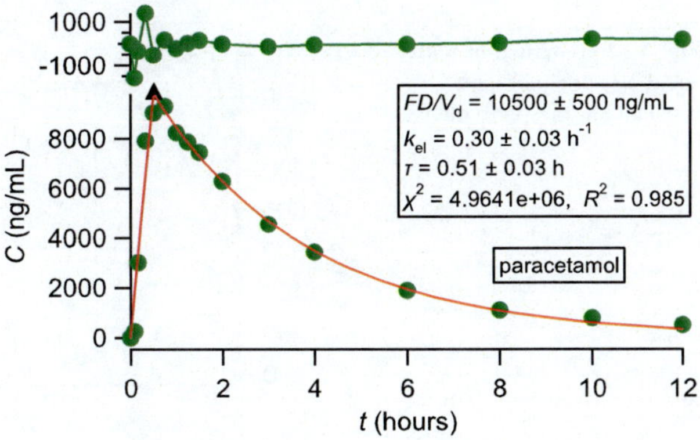

Fig. 4.31 One-compartment, two input fit to paracetamol data [5]

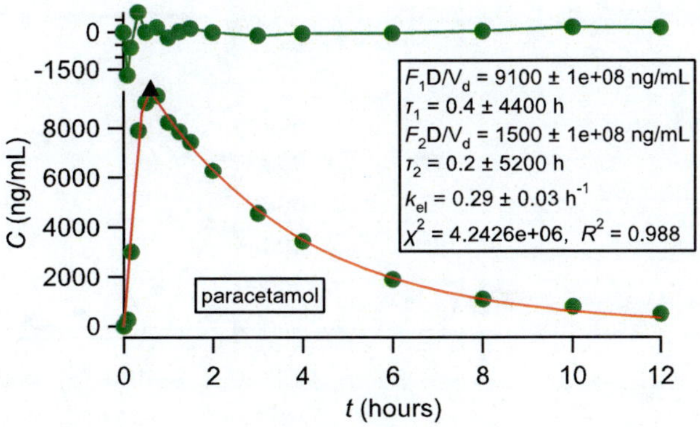

significant improvement, with the latter providing a closer fit (lower χ^2), although its overall quality is not better given the high parameter uncertainties.

Niraparib pharmacokinetic data [6] are analyzed next. The standard one-compartment model with first-order input and elimination kinetics does not describe in sufficient detail the data, Fig. 4.32. The one-compartment model with constant input rate and first-order elimination kinetics fares better, but still misses the correct data pattern, Fig. 4.33. The addition of a peripheral department improves the quality of the fit, Fig. 4.34. This is further improved by the use of two successive constant input rates. High correlations among the parameters yield large uncertainties which are reduced by constraining one of both τ_1 and τ_2, see Figs. 4.35, 4.36, and 4.37. Tables 4.9, 4.10, and 4.11 show the impact of holding one or two parameters constant at the optimum values on parameter uncertainties and parameter correlations.

A significantly larger number of data sets (not included here) have been analyzed in the same way. The reader is invited to use this software (provided in the electronic supplementary material) to analyze their data.

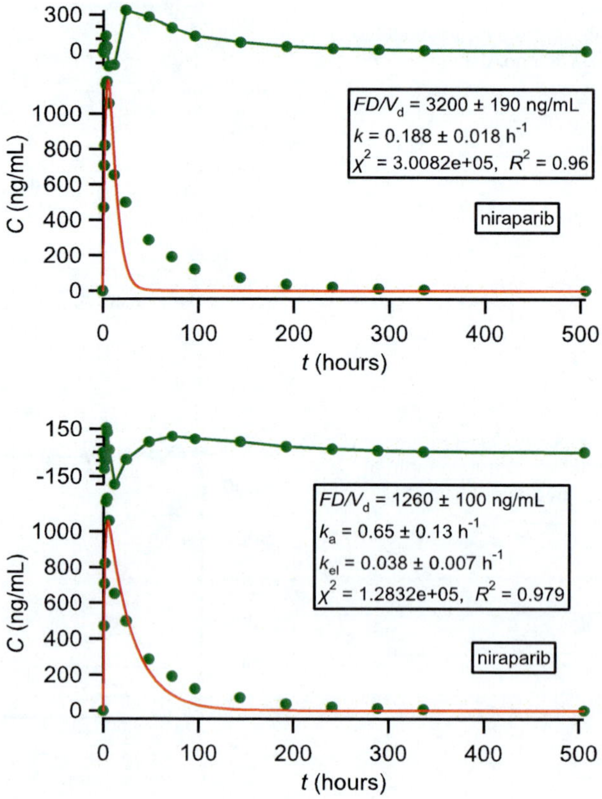

Fig. 4.32 Bateman fit with equal and unequal rate constants to niraparib data [6]

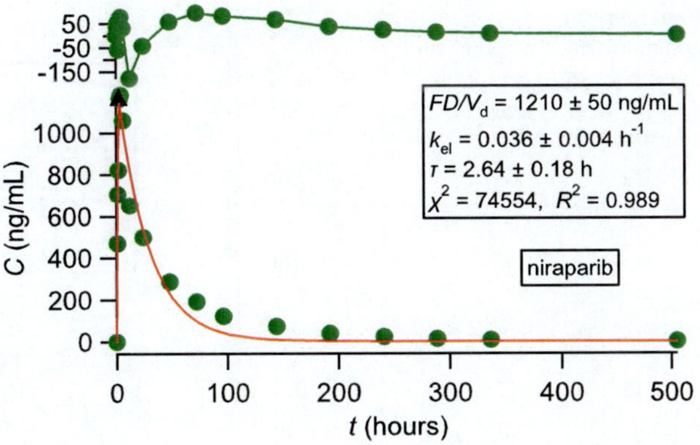

Fig. 4.33 One-compartment, one constant input rate fit to niraparib data [6]

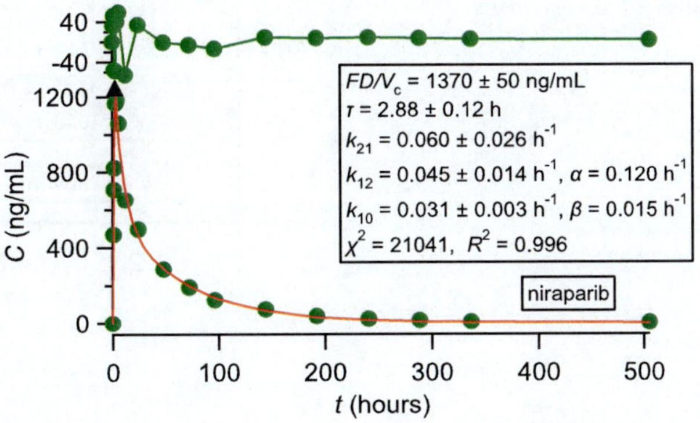

Fig. 4.34 Two-compartment, one constant input rate fit to niraparib data [6]

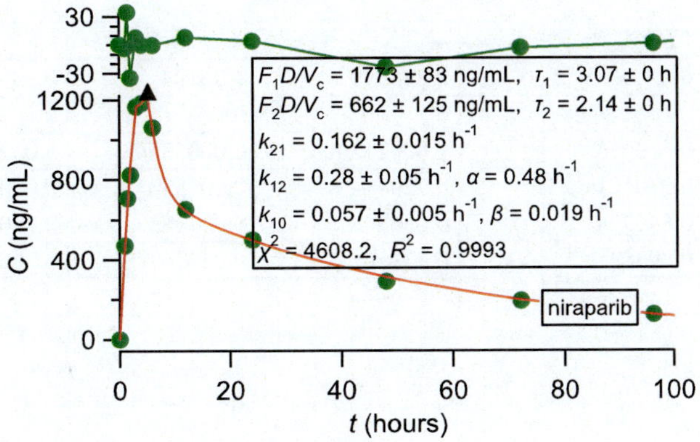

Fig. 4.35 Two-compartment, two constant input rates fit to niraparib data [6] with both τ_1 and τ_2 held constant

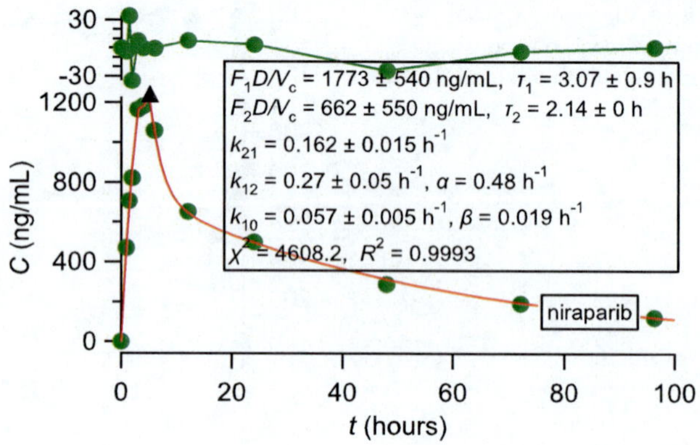

Fig. 4.36 Two-compartment, two constant input rates fit to niraparib data [6] with τ_2 held constant

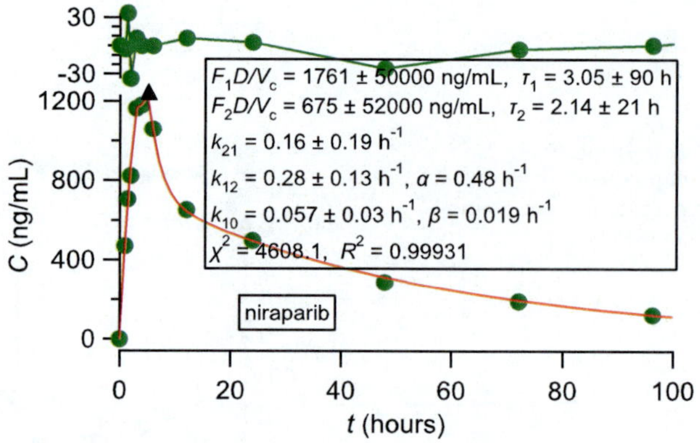

Fig. 4.37 Two-compartment, two constant input rates fit to niraparib data [6] with all parameters allowed to vary freely

Table 4.9 Variances (σ_i^2), covariances (σ_{ij}^2), and correlation coefficients (r_{ij}^2) among parameters for the results shown in Fig. 4.35. τ_1 and τ_2 have zero variance, covariances and correlation coefficients because they are held constant

F_1D/V_c	τ_1	k_{21}	k_{12}	k_{10}	F_2D/V_c	τ_2
6926.06	0	0.825519	4.06104	0.393644	9305.55	0
	0	0	0	0	0	0
0.679656		0.000213005	0.00053548	5.0529e-05	0.970432	0
0.970121		0.729424	0.0025301	0.000239035	5.89402	0
0.896974		0.656547	0.901183	2.78075e-05	0.59765	0
0.89389		0.531564	0.93676	0.906049	15646.9	0
						0

Table 4.10 Variances (σ_i^2), covariances (σ_{ij}^2), and correlation coefficients (r_{ij}^2) among parameters for the results shown in Fig. 4.36. Values related to τ_2 are not included because it was held constant

F_1D/V_c	τ_1	k_{21}	k_{12}	k_{10}	F_2D/V_c
291,270	480.401	1.41278	6.87691	0.49284	−271,697
0.98704	0.813284	0.000877882	0.0041958	0.00011255	−477.027
0.171969	0.0639493	0.000231717	0.000584644	5.48641e-05	0.536538
0.242444	0.0885236	0.730765	0.00276229	0.000259511	3.92397
0.166341	0.0227334	0.656524	0.89942	3.01381e-05	0.581424
−0.924149	−0.971018	0.0647034	0.137056	0.19442	296,749

4.4 Toward a Biopharmaceutics–Pharmacokinetic Classification System (BPCS)

The analysis of data presented above underlines the fact that the duration, τ, of the absorption process is a fundamental biopharmaceutical parameter of drug when administered as an immediate release formulation. The type of immediate release formulation can also have an impact on the τ estimate (see cyclosporine results, Figs. 4.17 and 4.23). For years and years, the absorption rate constant became the sole parameter for expressing quantitatively the rate of drug absorption in classical and population pharmacokinetic studies as explained in Chap. 2. The results presented in Figs. 4.12, 4.13, 4.14, 4.15, 4.16, 4.17, 4.18, 4.19, 4.20, 4.21, 4.22, 4.23, 4.24, 4.25, 4.26, 4.27, 4.28, 4.29, 4.30, 4.31, 4.32, 4.33, 4.34, 4.35, 4.36, and 4.37 clearly demonstrate the superiority of PBFTPK models over the classical first-order absorption models for the description of kinetic characteristics of drugs/formulations. Roughly, the more complex the absorption is the better is the performance of PBFTPK models compared to the Bateman equation (Eq. 1.2).

The current analysis relies on the FAT concept (Chap. 3) and allows the estimation of τ, which can characterize each drug/formulation given as an immediate release formulation [1]. This is so since τ is conceptually associated with the fundamental biopharmaceutical properties of solubility and permeability as shown in Chap. 3. Intuitively, drugs/immediate release formulations can be classified into (1) rapidly absorbing $\tau < 1.5$ h like paracetamol and borderline cyclosporine (Sandimmune Neoral) administered under fasted conditions; (2) medium absorbing $1.5 \leq \tau < 5$ h like ibuprofen, almotriptan [1], cyclosporine (Sandimmune Neoral) administered under fed conditions as well as cyclosporine (Sandimmune) administered under fasted conditions and niraparib; (3) slow absorbing $5 \leq \tau < 30$ h not observed in the data presented the data analyzed. For the first two categories, drug absorption takes place only in the small intestine, while for the third category, colon absorption is also operating. Several drugs/formulations exhibiting either selective regional permeability or solubility/ionization characteristics that lead to precipitation/re-dissolution comprise a fourth category characterized by a complex

Table 4.11 Variances (σ_i^2), covariances (σ_{ij}^2), and correlation coefficients (r_{ij}^2) among parameters for the results shown in Fig. 4.37

F_1D/V_c	τ_1	k_{21}	k_{12}	k_{10}	F_2D/V_c	τ_2
2.5285e9	4.37766e6	9519.64	6096.55	−1582.21	−2.6067e9	−1.0388e6
0.999998	7579.06	16.4798	10.5467	−2.74011	−4.513e6	−1798.44
0.996578	0.99648	0.036087	0.023560	−0.00590	−9811.91	−3.91068
0.911915	0.911198	0.932854	0.0176762	−0.00353	−6273.09	−2.50177
−0.98339	−0.983678	−0.97052	−0.82952	0.0010238	1632.34	0.650324
−0.99999	−0.999997	−0.99636	−0.910174	0.984106	2.68735e9	1.07091e6
−0.99996	−0.999968	−0.99650	−0.910855	0.98383	0.999967	426.782

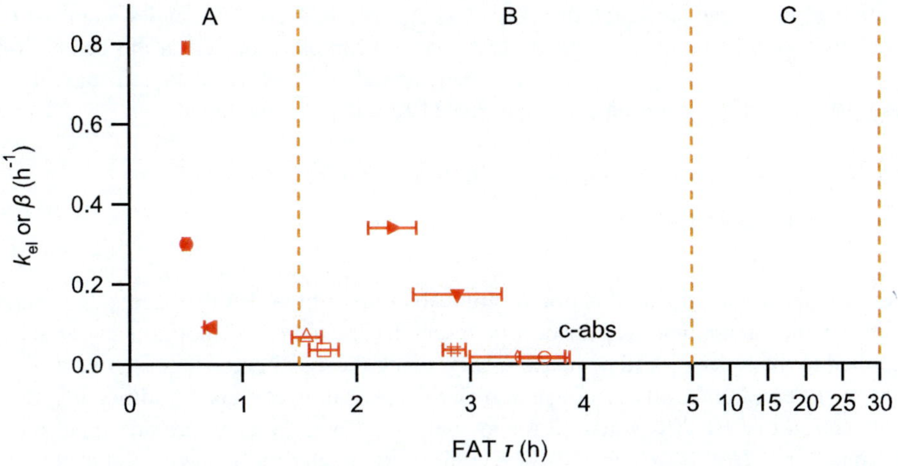

Fig. 4.38 Plot of elimination rate constant, k_{el} or β, estimates vs. finite absorption time (FAT, τ) estimates (±SD) [1] Key: paracetamol (●), cyclosporine (Sandimmune Neoral, fasted) (Δ), ibuprofen (►), almotriptan (▼), cyclosporine (Sandimmune Neoral fed) (□), cyclosporine (Sandimmune, fasted) (#), niraparib (◊), theophylline [7, 8] (◄), BMS-626529 drug [7, 9] (♦). Filled symbols correspond to k_{el} estimates (one-compartment model drugs), while empty symbols correspond to β estimates (two-compartment model drugs). The term c-abs next to cyclosporine (Sandimmune, fed) (o) administered under fed conditions, denotes complex absorption [1]

absorption profile like cyclosporine (Sandimmune) administered under fed conditions (see Fig. 4.23). Figure 4.38 shows the proposed three categories (A, B, and C) where a drug exhibiting complex absorption, denoted with c-abs, can also be classified in accord with its τ estimate [1]. All estimates for τ are coupled with the corresponding estimate for drug's elimination rate constant k_{el} or β for drugs obeying one or two-compartment model kinetics, respectively. Visual inspection of Fig. 4.38 reveals that the one-compartment model drugs paracetamol (Fig. 4.30) and theophylline [1, 2], which are biowaivers, are located in Class A close to the ordinate. This is in accord with their extensive absorption calculated from oral data, if one applies the one-compartment model methodology described for theophylline in [7]. This also applies to the BMS-626529 drug [7, 9]. All cyclosporine formulations, ibuprofen and almotriptan are classified in Class B [1].

Finally, it will be interesting to explore the classification presented in Fig. 4.38 in relation to other biopharmaceutical classifications [10–15]. In all examples analyzed the estimate for τ was found to be equal to t_{max}. Reliable estimates were derived for τ using our PBFTPK software, Figs. 4.14, 4.25, 4.28, 4.30, 4.35, and 4.36, since an adequate number of samples were available throughout the time course of drug in the body. For the one-compartment model drugs exhibiting one input rate like paracetamol, this finding, $\tau = t_{max}$, is a logical consequence of the FAT concept. On the contrary, estimates for τ were not found in the descending leg of the curves ($\tau > t_{max}$), which could be observed in other drugs (see Chap. 8). Although this is theoretically possible (Fig.4.4c, 4.5a, c, 4.7a, and 4.8c, e, f), the fitting results and the statistical measures presented in Figs. 4.14, 4.25, 4.28, 4.30, 4.35, and 4.36 provide conclusive evidence that $\tau = t_{max}$. However, the sampling design in the neighborhood of τ and the magnitude of the experimental error of the data can make the estimation of τ not possible using the PBFTPK software developed.

4.5 Future Work

As a matter of fact, the application of finite absorption time (FAT) concept can open new avenues in the oral drug absorption research. Thus, the FAT concept can be also applied to interspecies and pediatric scaling using the τ estimates for each one of the species or children/adult as a core parameter in the scaling exercise. Additionally, the application of PBFTPK software for re-analysis of oral data can provide input rate estimate(s) ($FD/\tau V_d$) which will be certainly associated with the rate-controlling parameter(s) of absorption, solubility, and/or permeability (see Chap. 3). Analysis of big oral data using machine learning techniques coupled with molecular descriptors can also elucidate critical factors of oral drug absorption phenomena. Besides, further applications of PBFTPK models to the following topics can be envisaged too: (1) development of models based on multiple oral drug administration; (2) construction of percent absorbed versus time plots and use in in vitro–in vivo correlations (IVIVC) under the prism of FAT concept; (3) extension/application of the modeling work to population (PK-PD) studies; and (4) coupling the PBFTPK modes with pharmacodynamic models. These applications (1–4) can be also considered in the light of non-linear (Michaelis–Menten) kinetics. All above, if coupled with the implications of finite absorption time models on bioavailability/ bioequivalence issues (Chaps. 7 and 8), point to a new era in the scientific and regulatory aspects of oral drug absorption.

References

1. Chryssafidis P, Tsekouras AA, Macheras P (2022) Re-writing oral pharmacokinetics using physiologically based finite time pharmacokinetic (PBFTPK) models. Pharm Res 39:691–701. https://doi.org/10.1007/s11095-022-03230-0
2. Dokoumetzidis A, Macheras P (1998) Investigation of absorption kinetics by the phase plane method. Pharm Res 15:1262–1269. https://doi.org/10.1023/A:1011952227079
3. Kesisoglou F, Mitra A (2015) Application of absorption modeling in rational design of drug product under quality-by-design paradigm. AAPS J 17:1224–1236. https://doi.org/10.1208/s12248-015-9781-1
4. Mueller EA, Kovarik JM, van Bree JB, Grevel J, Lucker PW, Kutz K (1994) Influence of a fat-rich meal on the pharmacokinetics of a new oral formulation of cyclosporine in a crossover comparison with the market formulation. Pharm Res 11:151–155. https://doi.org/10.1023/a:1018922517162
5. Atkinson HC, Stanescu I, Frampton C, Salem II, Beasleyr CPH, Robson R (2015) Pharmacokinetics and bioavailability of a fixed-dose combination of ibuprofen and paracetamol after

intravenous and oral administration. Clin Drug Investig 35:625–632. https://doi.org/10.1007/s40261-015-0320-8

6. van Andel L, Rosing H, Zhang Z, Hughes L, Kansra V, Sanghvi M, Tibben MM, Gebretensae A, Schellens JHM, Beijnen JH (2018) Determination of the absolute oral bioavailability of niraparib by simultaneous administration of a 14C-microtracer and therapeutic dose in cancer patients. Cancer Chemoth Pharmacol 81:39–46. https://doi.org/10.1007/s00280-017-3455-x
7. Chryssafidis P, Tsekouras AA, Macheras P (2021) Revising pharmacokinetics of oral drug absorption: II Bioavailability-bioequivalence considerations. Pharm Res 38:1345–1356. https://doi.org/10.1007/s11095-021-03078-w
8. Meyer MC, Jarvi EJ, Straughn AB, Pelsor FR, Williams RL, Shah VP (1999) Bioequivalence of immediate-release theophylline capsules. Biopharm Drug Disp 20:417–419. https://doi.org/10.1002/1099-081X(199912)20:9<417::AID-BDD205>3.0.CO;2-W
9. Brown J, Chien C, Timmins P, Dennis A, Doll W, Sandefer E, Page R, Nettles RE, Zhu L, Grasela D (2013) Compartmental absorption modeling and site of absorption studies to determine feasibility of an extended-release formulation of an HIV-1 attachment inhibitor phosphate Ester prodrug. J Pharm Sci 102:1742–1751. https://doi.org/10.1002/jps.23476
10. Amidon GL, Lennernas H, Shah VP, Crison JR (1995) A theoretical basis for a biopharmaceutic drug classification: the correlation of in vitro drug product dissolution and in vivo bioavailability. Pharm Res 12:413–420. https://doi.org/10.1023/A:1016212804288
11. Wu CY, Benet LZ (2005) Predicting drug disposition via application of BCS: transport/absorption/ elimination interplay and development of a biopharmaceutics drug disposition classification system. Pharm Res 22:11–23. https://doi.org/10.1007/s11095-004-9004-4
12. Macheras P, Karalis V (2014) A non-binary biopharmaceutical classification of drugs: the ΑΒΓ system. Int J Pharm 464:85–90. https://doi.org/10.1016/j.ijpharm.2014.01.022
13. Rinaki E, Valsami G, Macheras P (2003) Quantitative biopharmaceutics classification system: the central role of dose/solubility ratio. Pharm Res 20:1917–1925. https://doi.org/10.1023/B:PHAM.0000008037.57884.11
14. Charkoftaki G, Dokoumetzidis A, Valsami G, Macheras P (2012) Elucidating the role of dose in the biopharmaceutics classification of drugs: the concepts of critical dose, effective in vivo solubility, and dose-dependent BCS. Pharm Res 29:3188–3198. https://doi.org/10.1007/s11095-012-0815-4
15. Macheras P, Iliadis A, Melagraki G (2018) A reaction limited in vivo dissolution model for the study of drug absorption: towards a new paradigm for the biopharmaceutic classification of drugs. Eur J Pharm Sci 117:98–106. https://doi.org/10.1016/j.ejps.2018.02.003

Part II
Bioavailability-Bioequivalence

Chapter 5
History of the Bioavailability–Bioequivalence Concepts

Variability is the law of life, and as no two faces are the same, so no two bodies are alike, and no individuals react alike and behave alike under the abnormal conditions which we know as disease.
Sir William Osler (1849–1919)

Contents

Abstract This chapter reviews the first experimental observations associated with the "bioavailability problems" encountered in the 1960s and 1970s. The introduction of the bioavailability concept by the US Food and Drug Administration in 1977 was a logical consequence for the protection of public health.

Keywords Bioavailability · Bioequivalence · Rate of absorption · C_{max}

In *Metaphysics* A.1, Aristotle says that "everyone takes what is called 'wisdom' (*sophia*) to be concerned with the primary causes (*aitia*) and the starting-points (or principles, *archai*)" [1]. In a similar approach, this chapter focuses on the history of the thought of the predecessors in drug action prior to 1977 looking at the principles held in pharmacology, pharmacodynamics, biopharmaceutics, and pharmacokinetics at that time and the causes, which led to the establishment of the

P. Macheras, A. A. Tsekouras, *Revising Oral Pharmacokinetics, Bioavailability and Bioequivalence Based on the Finite Absorption Time Concept*,
https://doi.org/10.1007/978-3-031-20025-0_5

bioavailability concept. Hence, we are reviewing in this chapter the scientific background in relation to the adoption of bioavailability by the FDA on January 7, 1977 [2] for regulatory purposes. The major scientific and regulatory decisions/changes/advances took place in the USA during the decades 1950s, 1960s, and 1970s.

5.1 Early 1900s-Today: Variations in Response to Xenobiotics

Sir William Osler (1849–1919) was a Canadian physician and one of the four founding professors of Johns Hopkins Hospital. Osler's "Principles and Practice" became the bible for generations of medical students and practitioners. He will forever be immortalized in the medical world through the groundbreaking changes to medical training and education he implemented. He is considered the father of modern medicine; he is also famous for his definition of variability, "Variability is the *law* of life, and as no two faces are the same, so no two bodies are alike, and no individuals react alike and behave alike under the abnormal conditions which we know as disease" [3, 4].

This is a remarkable definition pointed out a century ago; it fully substantiates the modern definition of personalized medicine: "the right patient population, the right drug, the right dose, the right indication, and administration at the right time." In other words, the traditional methods of medicine are not always the best for all patients.

5.2 1950s: The First Pharmacokinetic–Pharmacodynamic Correlations

The basic reason for measuring drug concentrations in blood (plasma or serum) is the correlation of the drug's blood levels with the pharmacodynamics effect or the efficacy of the drug. Although this sounds logical and well founded in our days, this realization was started in the 1950s and 1960s. The most notable example is the determination of the critical serum concentration of quinidine for the prevention of attacks of paroxysmal ventricular tachycardia [5], Fig. 5.1.

Another notable example is the correlation of blood alcohol concentration (BAC) with the successive stages (subliminal intoxication, euphoria, excitement, confusion, stupor, coma, death) of alcohol intoxication. Based on relevant studies [6] in the 1970s, the legal BAC limit for a driver was assigned to 0.08, namely, ethanol should be less than 0.08 g in 100 mL of blood.

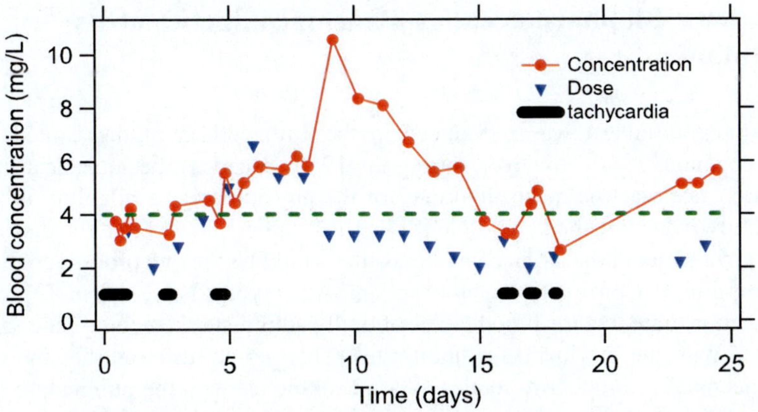

Fig. 5.1 Quinidine serum levels (mg/L) and attacks of paroxysmal ventricular tachycardia as a function of time for a patient with coronary heart disease [5]. The critical quinidine serum level is 4 mg/L above of which attacks of paroxysmal ventricular tachycardia are prevented

5.3 1953: The Publication of Dost's First Pharmacokinetics Book in History

According to Wikipedia, Friedrich Hartmut Dost was born on July 11, 1910, in Dresden and died on November 02, 1985 in Giessen. He studied medicine in Rostock, Freiburg, Innsbruck, and Leipzig. In 1960 became a Professor and director of the pediatric hospital of the Justus Liebig University of Giessen. He introduced the term pharmacokinetics in 1953 in his text, Der Blutspiegel-Kinetic der Konzentrationsabläufe in der Kreislaufflüssigkeit [7].

This outstanding book for its time fully covered the so-called one-compartment open model with its various forms of input. We recall here Eq. (5.1) from Dost's book and Chap. 1 for the purposes of the present chapter. Eq. (5.1) describes the concentration of drugs in blood as a function of time for the linear one-compartment model with first-order absorption and elimination:

$$C_{\mathrm{b}}(t) = \frac{FDk_{\mathrm{a}}}{V_{\mathrm{d}}(k_{\mathrm{a}} - k_{\mathrm{el}})}\left(e^{-k_{\mathrm{el}}t} - e^{-k_{\mathrm{a}}t}\right) \tag{5.1}$$

where F is the fraction of dose, D absorbed, V_{d} is the volume of distribution, k_{a} is the first-order rate constant and k_{el} is the elimination rate constant. All fundamental parameters associated with bioavailability, namely $[\mathrm{AUC}]_0^{\infty}$, $C_{\max}$ and $t_{\max}$ (see Eqs. 1.4, 1.5, 1.6, respectively) are derived from Eq. (5.1) as explained in Chap. 1.

5.4 1960s: Biopharmaceutics–Pharmacokinetics at Its Infancy

The disintegration test was introduced in the 14th edition of the United States Pharmacopeia in 1950. However, during the 1950s decade, it became clear that the dissolution rate was the most important for the physiological availability of drugs. Indeed, Edwards published an article [8] in 1951 and postulated that the dissolution of an aspirin tablet in the stomach and intestine would be the rate process controlling the absorption of aspirin into the bloodstream. A few years later, Nelson [9] was the first to demonstrate that the blood levels of orally administered theophylline salts are correlated with the in vitro dissolution rates. This period also signifies the formal commencement of biopharmaceutics–pharmacokinetics with the publication of two review articles in the *Journal of Pharmaceutical Sciences* by the pioneer scientists Eino Nelson [10] and John Wagner [11] in 1961. These articles summarized the 1960s state of the art on the kinetics of drug Absorption, Distribution, Metabolism, and Excretion (ADME).

5.5 "The Bioavailability Problem"

As mentioned above, variability in drug response was always associated with the patient in accordance with Sir William Osler's variability principle [3, 4]. In the late 1960s, however, it was realized that a variable or poor response to a therapeutic agent may not have its origin in the patient; it may be due to a formulation defect in the drug product administered [12], the so-called "bioavailability problem." In his review article [13] on the "history of biopharmaceutics in FDA 1968–1993" Jerome Skelly reveals that the "bioavailability problem" term was used during this early period of the publication of the pertinent bioavailability papers. These papers demonstrated that differences in product formulation could result in large differences in drug response. Several examples can be quoted: (1) the lack of clinical effect for two prednisone products [14, 15]; (2) significantly lower blood levels and hypoglycemic effect were observed as a result of a slight change in the formulation of an experimental tolbutamide preparation [16]; (3) significant differences in the bioavailability between different brands of sodium diphenylhydantoin, chloramphenicol, and sulfisoxazole were reported [17] by Martin et al. in 1968; and (4) greater than 20% difference in peak concentration and area under the serum concentration–time curve for three ampicillin products were also found [18] in 1972; it should be added, the tremendous differences found in the concentration–time profiles found [19] in the generic chloramphenicol products of the US market, Fig. 5.2.

However, the most serious bioavailability problems were encountered with digoxin [12, 20] in the UK and the USA in 1971 and phenytoin in Australia and New Zealand in 1968. A detailed description of the digoxin bioavailability problems in the USA is given by J. Skelly in his review [13]. Extensive dissolution studies

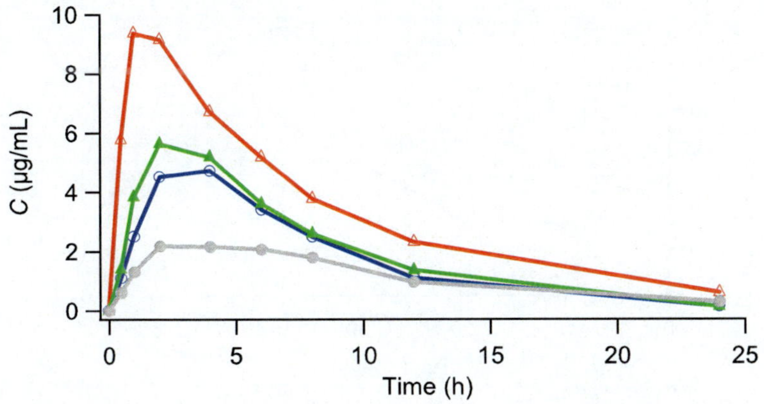

Fig. 5.2 Mean plasma levels of chloramphenicol in ten human subjects following 0.5 g oral doses in various formulations [19]

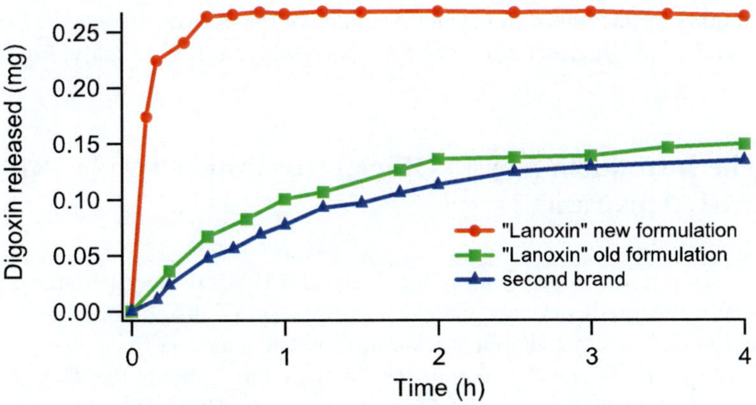

Fig. 5.3 Dissolution profiles of three different formulations of digoxin, exhibiting large differences [21]

were carried out with the 0.25-mg digoxin tablets available in the 1972 North American marketplace [13]. An additional example [21] demonstrating the large differences found in the dissolution profiles of three digoxin formulations is shown in Fig. 5.3. All dissolution digoxin studies unequivocally demonstrated that all differences in the dissolution profiles of the digoxin products (either lot-to-lot or among brands), are reflected on the bioequivalence differences observed.

The bioavailability problem with phenytoin was even more important since phenytoin toxicity occurred in a large number of patients. In fact, the manufacturer replaced the excipient calcium sulfate with lactose in immediate release phenytoin tablets; this replacement caused the phenytoin intoxication [22]. The more hydrophilic lactose enhanced the dissolution rate of phenytoin leading to higher

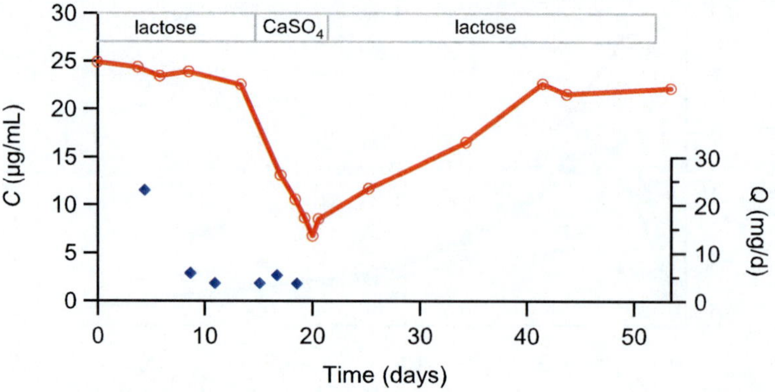

Fig. 5.4 Blood phenytoin concentrations in a patient taking phenytoin (400 mg/day), with excipients respectively as shown (lactose, calcium sulfate, lactose). Blue diamonds represent daily fecal excretion of phenytoin when measured [22]

concentrations of phenytoin in plasma, well above the narrow therapeutic range of 10–20 μg/mL. This interpretation was confirmed with an in vivo study, Fig. 5.4.

5.6 The Introduction of an Official Dissolution Test in 1970 (USP Apparatus 1)

The dissolution studies carried out in the 1950s and 1960s dealing with the effect of pharmaceutical ingredients and processes on the rate of drug dissolution coupled with the dissolution bioavailability relationships delineated in Sect. 5.5, prompted FDA to adopt an official dissolution test in 6 monographs of the *United States Pharmacopeia* (*USP*) and *National Formulary* (*NF*) in 1970. This decision underlines the prodigious value of dissolution testing as a tool for quality control. Since then equivalence in dissolution behavior is always under the prism of both the bioavailability and the quality control considerations. Although the basket-stirred-flask test (USP apparatus 1) was adopted as an official dissolution test in only 6 monographs of the *United States Pharmacopeia* (*USP*) and *National Formulary* (*NF*) in 1970, tremendous developments took place in the ensuing years with an explosion not only in the number of monographs of immediate release dosage forms but also in the number of modified and delayed dosage forms [23].

5.7 FDA (1977): Bioavailability Is the Rate and Extent to Which the Active Ingredient or Active Moiety Is Absorbed from a Drug Product and Becomes Available at the Site of Action

The adoption of the bioavailability concept by FDA in 1977 is the logical consequence of in vitro and in vivo experimental observations published in the literature and described in Sect. 5.5. Figure 5.5 presents in chronological order the relevant developments in this field of research.

The definition of bioavailability adopted by FDA is very similar to the definition quoted in a booklet published [24] by the American Association of Pharmaceutical Sciences in 1972: "Bioavailability is a term used to indicate measurement of both the relative amount of an administered drug that reaches the general circulation and the rate at which this occurs. In the context of this definition, general circulation refers primarily to the venous blood (excepting the hepatic portal blood during the absorptive phase) and arterial blood which carry the drug to the tissues." In addition, this document [24] under the subtitle *"Measurement of the extent of bioavailability"* explicitly refers to (1) "measurement of the area under the blood, plasma or serum concentration-time curve (AUC)" and (2) "one must include an estimate of the AUC beyond the last data point." These extracts from [24] clearly show that the use of AUC as a metric for the measurement of the extent of absorption was established several years before the official definition of bioavailability from FDA on January 7, 1977.

However, the section of the booklet [24] entitled *"Measurement of the rate of* bioavailability" does not mention C_{max} at all. The only indirect reference to C_{max} is made in the "Comparative peak time analysis" subsection [24] whereas one can read: "When the same drug is administered in different dosage forms, the time the blood

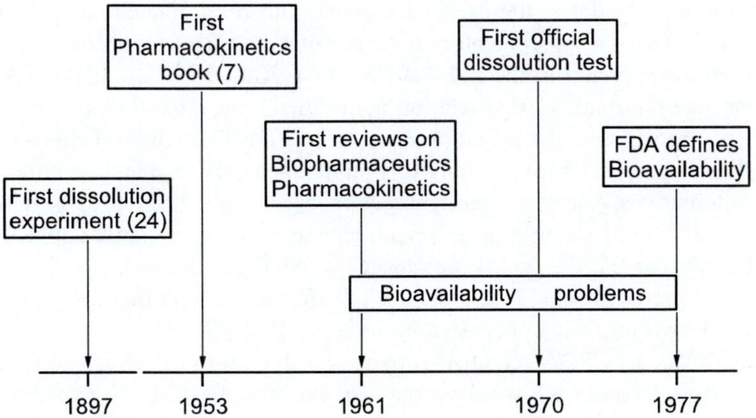

Fig. 5.5 Major scientific and regulatory developments prior to the adoption of the bioavailability concept by FDA in 1977

concentration curve reaches its peak in a given subject is related to the rate of absorption." It can be concluded that no specific metric for the rate of drug absorption was proposed up to 1972 [24]. According to the committee members (L.W. Ditter, W. A. Cressman, S.A. Kaplan, S. Riegelman, and J. Wagner) who wrote the Guidelines for biopharmaceutical studies in man [24]: "Assessment of the rate of bioavailability is one of the most difficult problems encountered in bioavailability studies."

We quote here two bioavailability studies [25, 26] from this early period focusing on the metrics used for the assessment of bioavailability–bioequivalence. Both studies were published prior to the definition of bioavailability by FDA in 1977. The first study [25], published in 1973, shows that neither AUC nor C_{max} reflects the extent of digoxin absorption if contrasted with the 5-day urine digoxin excretion data. The second study [26], published in 1975, clearly shows that the area under the blood concentration, time of truncated curves of ten drugs are equally well metrics, compared with the untruncated areas, for the assessment of bioavailability. Both studies [25, 26] show the ambiguity associated with the reliability/robustness of the relevant metrics; these studies [25, 26] will be reconsidered in Chap. 8.

5.7.1 *A Persisting Problem: The Use of* C_{max} *as a Metric of Rate of Absorption*

The origin, e.g., a document for the use of the term "rate" in the FDA definition: "*Bioavailability is the rate and extent*..." cannot be identified. Perhaps, its use was based on the predominant role of the "rate of dissolution" in the in vitro studies and its connection with the drug's bioavailability, e.g. [9]. Many of the in vitro studies in 1950s, 1960s, and 1970s were revolving around the "rate of dissolution" term; this is also the case for the first ever dissolution study published in 1897 [27], Fig. 5.5.

From these early dates, the use of the maximum drug concentration in blood, C_{max} is being used as a sole regulatory indicator of a drug's rate of absorption. Its use can be partly interpreted in the light of Eq. (1.5). Although C_{max} is related to the absorption rate constant, k_a, this relation is not linear but most importantly C_{max} is also related linearly with the extent of absorption (*FD*). These hybrid characteristics of C_{max} have been criticized extensively and repeatedly in the literature. Thus, various alternatives (methods and indirect metrics) have been proposed for the assessment of rate of absorption in bioequivalence studies, namely, application of moment analysis [28], the use of the ratio C_{max}/AUC instead of C_{max} [29–31], the estimation of partial areas [32, 33], the use of different indirect metrics [34–36] and an analysis based on early concentration–time profiles [37, 38].

Twenty years ago, FDA scientists proposed a shift from the classical extent and rate of absorption concepts toward measures of exposure [39], arguing that the "*Rate of absorption is not only difficult to measure but also bears little clinical relevance.*" Almost 30 years ago, Tucker et al. [40] suggested "*that the ambiguity in the rationale for bioequivalence testing would be removed if the term "rate" was deleted from its definition"* since *"there is no rate parameter which allows products to be compared for both pharmaceutical quality (actual release rate) and clinical safety and efficacy."* Despite of the mentioned shortcomings and criticism, $C_{\max}$ is always being used as a rate parameter in all bioequivalence guidelines worldwide.

Chapters 7 and 8 focus on bioavailability and bioequivalence issues under the prism of the finite absorption time concept developed recently [41–43]. In this context, $C_{\max}$, which is either equal to or higher than the drug concentration at the end of the absorption process is reinterpreted.

References

1. Aristotle. Metaphysics A.1
2. FDA. On January 7, 1977, FDA issued final regulations in part 320 (21 CFR 320) establishing definitions and requirements for BA and BE studies (42 FR 1624)
3. Marrer E, Baty F, Kehren J, Chibout S-D, Brutsche M (2006) Past, present and future of gene expression-tailored therapy for lung cancer. Per Med 3:165–175. https://doi.org/10.2217/17410541.3.2.165
4. Hong K-W, Bermseok O (2010) Overview of personalized medicine in the disease genomic era. BMB Rep 43:643–648. https://doi.org/10.5483/BMBRep.2010.43.10.643
5. Sokolow M, Edgar L (1950) Blood quinidine concentrations as a guide in the treatment of cardiac arrhythmias. Circulation 1:576–592. https://doi.org/10.1161/01.CIR.1.4.576
6. Sidell FR, Pless JE (1971) Ethyl alcohol: blood levels and performance decrements after oral administration to man. Psychopharmacologia 19:246–261. https://doi.org/10.1007/BF00401941
7. Dost FH (1953) Der Blutspiegel. Kinetik der Konzentrationsabläufe in der Kreislaufflüssigkeit. Thieme, Leipzig
8. Edwards LJ (1951) The dissolution and diffusion of aspirin in aqueous media. Trans Faraday Soc 47:1191–1210. https://doi.org/10.1039/TF9514701191
9. Nelson E (1957) Solution rate of theophylline salts and effects from oral administration. J Am Pharm Assoc 46:607–614. https://doi.org/10.1002/jps.3030461012
10. Nelson E (1961) Kinetics of drug absorption, distribution, metabolism and excretion. J Pharm Sci 50:181–192. https://doi.org/10.1002/jps.2600500302
11. Wagner JG (1961) Biopharmaceutics: absorption aspects. J Pharm Sci 50:359–387. https://doi.org/10.1002/jps.2600500502
12. Vitti TG, Banes D, Byers TE, Bioavailability of digoxin (1971) N Engl J Med 285:1433. https://doi.org/10.1056/nejm197112162852512
13. Skelly JP (2010) The history of biopharmaceuitcs in Food and Drug Administration 1968-1993. AAPS J 12:44–50. https://doi.org/10.1208/s12248-009-9154-8
14. Campagna FA, Cureton G, Mirigian RA, Nelson E (1963) Inactive prednisone tablets USP XVI. J Pharm Sci 52:605–660. https://doi.org/10.1002/jps.2600520626
15. Levy G, Hall NA, Nelson E (1964) Studies on inactive prednisone tablets USP XVI. Am J Hosp Pharm 21:402. https://doi.org/10.1093/ajhp/21.9.402
16. Varley AB (1968) The generic inequivalence of drugs. JAMA 206:1745–1748. https://doi.org/10.1111/j.1600-0773.1971.tb03308.x
17. Martin CM, Rubin M, O'Malley WE, Garagusi VF, McCauley CE (1968) Brand, generic drugs differ in man. JAMA 205(9):23. https://doi.org/10.1001/jama.1968.03140350005003

18. MacLeod C, Rabin H, Ruedy J, Caron M, Zarowny D, Davies R (1972) Comparative bioavailability of three brands of ampicillin. Can Med Assoc J 107:203–209. www.cmaj.ca/content/107/3/203
19. Glazko AJ, Kinkel AW, Alegnani WC, Holmes EL (1968) An evaluation of the absorption characteristics of different chloramphenicol preparations in normal human subjects. Clin Pharmacol Ther 9:472–548. https://doi.org/10.1002/cpt196894472
20. Lindenbaum J, Mellow MH, Blackstone MO, Butler VP Jr (1971) Variation in biologic availability of digoxin from four preparations. N Engl J Med 285:1344–1347. https://doi.org/10.1056/NEJM197112092852403
21. Fraser EJ, Leach RH, Poston JW (1972) Bioavailability of digoxin. Lancet 2:541. https://doi.org/10.1016/s0140-6736(72)91936-8
22. Tyrer JH, Eadie MJ, Sutherland JM, Hooper WD (1970) Outbreak of anticonvulsant intoxication in an Australian city. Br Med J 4:271–273. https://doi.org/10.1136/bmj.4.5730.271
23. Dokoumetzidis A, Macheras P (2006) A century of dissolution research: from Noyes and Whitney to the biopharmaceutics classification system. Int J Pharm 321:1–11. https://doi.org/10.1016/j.ijpharm.2006.07.011
24. Guidelines for biopharmaceutical studies in man. American Association of Pharmaceutical Sciences, Academy of pharmaceutical sciences, Washington, DC, February 1972
25. Sanchez N, Sheiner LB, Halkin H, Melmon KL (1973) Pharmacokinetics of digoxin: interpreting bioavailability. Br Med J 4:132. https://doi.org/10.1136/bmj.4.5885.132
26. Lovering EG, McGilveray IJ, McMillan I, Tostowaryk W (1975) Comparative bioavailabilities from truncated blood level curves. J Pharm Sci 64:1521–1524. https://doi.org/10.1002/jps.2600640921
27. Noyes AA, Whitney WR (1897) The rate of solution of solid substances in their own solutions. J Am Chem Soc 19:930–934. https://doi.org/10.1021/ja02086a003
28. Jackson AJ, Chen ML (1987) Application of moment analysis in assessing rates of absorption for bioequivalency studies. J Pharm Sci 76:6–9. https://doi.org/10.1002/jps.2600760103
29. Endrenyi L, Fritsch S, Yan W (1991) Cmax/AUC is a clearer measure than Cmax for absorption rates in investigations of bioequivalence. Int J Clin Pharmacol Ther Toxicol 29:394–399
30. Endrenyi L, Yan W (1993) Variation of Cmax and Cmax/AUC in investigations of bioequivalence. Int J Clin Pharmacol Ther Toxicol 31:184–189
31. Tozer TN, Hauck WW (1997) Cmax/AUC, a commentary. Pharm Res 14:967–968. https://doi.org/10.1023/a:1012128623213
32. Chen ML (1992) An alternative approach for assessment of rate of absorption in bioequivalence studies. Pharm Res 9:1380–1385. https://doi.org/10.1023/a:1015842425553
33. Macheras P, Symillides M, Reppas C (1994) The cutoff time point of the partial area method for assessment of rate of absorption in bioequivalence studies. Pharm Res 11:831–834. https://doi.org/10.1023/a:1018921622981
34. Lacey LF, Keene ON, Duquesnoy C, Bye A (1994) Evaluation of different indirect measures of rate of drug absorption in comparative pharmacokinetic studies. J Pharm Sci 83:212–215. https://doi.org/10.1002/jps.2600830219
35. Rostami-Hodjegan A, Jackson PR, Tucker GT (1994) Sensitivity of indirect metrics for assessing "rate" in bioequivalence studies—moving the "goalposts" or changing the "game". J Pharm Sci 83:1554–1557. https://doi.org/10.1002/jps.2600831107
36. Reppas C, Lacey LF, Keene ON, Macheras P, Bye A (1995) Evaluation of different metrics as indirect measures of rate of drug absorption from extended release dosage forms at steady-state. Pharm Res 12:103–107. https://doi.org/10.1023/a:1016246922519
37. Endrenyi L, Csizmadia F, Tothfalusi L, Chen ML (1998) Metrics comparing simulated early concentration profiles for the determination of bioequivalence. Pharm Res 15:1292–1299. https://doi.org/10.1023/a:1011912512966
38. Macheras P, Symillides M, Reppas C (1996) An improved intercept method for the assessment of absorption rate in bioequivalence studies. Pharm Res 13:1755–1758. https://doi.org/10.1023/a:1016421630290

39. Chen ML, Lesko L, Williams RL (2001) Measures of exposure versus measures of rate and extent of absorption. Clin Pharmacokinet 40:565–572. https://doi.org/10.2165/00003088-200140080-00001
40. Tucker GT, Rostami-Hodjegan A, Jackson PR (1995) Bioequivalence-a measure of therapeutic equivalence? In: Blume H, Midha K (eds) Bio-international 2, bioavailability, bioequivalence and pharmacokinetic studies. Medpharma Scientific publishers, Stuttgart, pp 35–43
41. Macheras P, Chryssafidis P (2020) Revising pharmacokinetics of oral drug absorption: I models based on biopharmaceutical/physiological and finite absorption time concepts. Pharm Res 37: 187. https://doi.org/10.1007/s11095-020-02894-w
42. Chryssafidis P, Tsekouras AA, Macheras P (2021) Revising pharmacokinetics of oral drug absorption: II Bioavailability-bioequivalence considerations. Pharm Res 38:1345–1356. https://doi.org/10.1007/s11095-021-03078-w
43. Tsekouras AA, Macheras P (2021) Re-examining digoxin bioavailability after half a century: time for changes in the bioavailability concepts. Pharm Res 38:1655–1638. https://doi.org/10.1007/s11095-021-03121-w

Chapter 6
Therapeutic Equivalence Based on Bioequivalence Studies

Time is the wisest of all things that are, for it brings everything to light.
Thales of Miletus (624–546 BC)

Contents

Abstract The basics of bioequivalence criteria for the assessment of therapeutic equivalence of generics versus the innovator's products are provided. The problem of highly variable drugs is emphasized as well.

Keywords Bioequivalence · Therapeutic equivalence · Bioequivalence criteria · Highly variable drugs

6.1 From 1977 to Now: Bioequivalence Criteria and Issues

Thousands of bioequivalence studies were performed worldwide upon the adoption of the bioavailability concept by FDA in 1977 until today. In all these studies, the crux of the matter is the statistical outcome of the decision for or against a determination of the therapeutic equivalence. Typically, therapeutic equivalence is declared when two products are bioequivalent. Initially, the decision rules for the approval of generics in the 1970s were based on mean AUC and $C_{\max}$ data of the formulations. The mean values of these parameters for the generic product (test) had to be within $\pm 20\%$ of those of the reference (innovator) product. In the late 1970s, the so called 75/75 or 75/75-125 rule was introduced to account for the individual variability in

P. Macheras, A. A. Tsekouras, *Revising Oral Pharmacokinetics, Bioavailability and Bioequivalence Based on the Finite Absorption Time Concept*,
https://doi.org/10.1007/978-3-031-20025-0_6

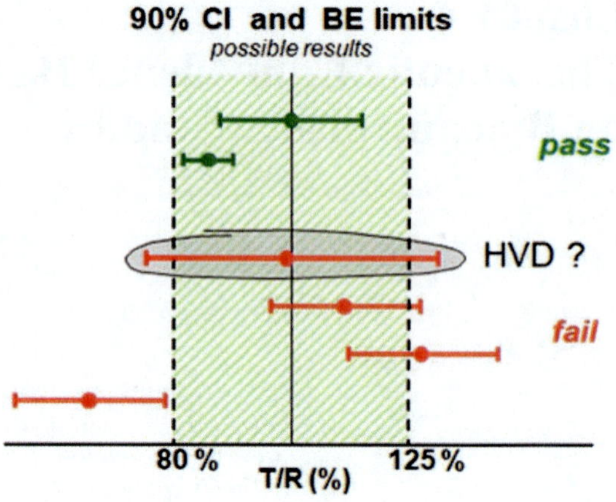

Fig. 6.1 Schematic depicting the 90% CI, the bioequivalence limits (80, 125%) for six studies (formulations of different drugs) exhibiting different *T/R*(%) ratios and different variability. The upper limit of 125% corresponds to the reciprocal of 80, namely (1/80)100%. HVD denotes a highly variable drug or drug product(s)

the rate and extent of absorption. According to this rule, the ratio of the test/reference for AUC and C_{max} must be within 0.75–1.25 for at least 75% of the subjects.

In 1986, all the above approaches were discontinued and the "*two-one-sided tests* (TOST)" procedure was applied [1]. This is also called *"the 90% confidence interval (CI) approach."* The methodology relies on an FDA survey of physicians who suggested that for most drugs, a difference of up to 20% in dose between two treatments would have no clinical significance. In other words, two formulations whose rate and extent of absorption differ by −20%/+25% or less are generally considered bioequivalent. For a number of reasons (pharmacokinetic models are multiplicative rather than additive, and bioequivalence comparisons are expressed as ratios rather than differences) the AUC and C_{max} values are logarithmically (ln) transformed. Accordingly, the calculated 90%CI from the ln transformed mean values of AUC and C_{max} should lie between 80 and 125%, which are called the bioequivalence limits. The upper or lower limits for the 90%CI are as follows:

$$90\%\mathrm{CI} = \exp\left[(m_T - m_R) \pm t_{0.05,N-2}\, s\sqrt{2/N}\right] \tag{6.1}$$

where m_T, m_R are the observed Test (T) and Reference means (R) of the log-transformed measures (AUC and C_{max}), N is the total number of subjects participating in the crossover study, $t_{0.05,\ N\text{-}2}$ is the point that isolates probability of 0.05 in the upper tail of the Student's t distribution with N-2 degrees of freedom and s is the square root of error mean square from the crossover design analysis of variance (ANOVA). The estimate for s is derived from the residual variance of the ANOVA and represents what we call intra-subject variability. The factor $s(2/N)^{1/2}$ corresponds to the standard error of the estimate. All above apply to a balanced crossover design with an equal number of subjects in each treatment-administration sequence while there are no missing observations from any subject.

Figure 6.1 shows the 90% CI vis-a-vis the bioequivalence limits for six pairs of different formulations which exhibit different *T/R* (%) means and different variability. The top one study shows an ideal 100% *T*/*R* value and relatively high variability; however, the 90%CI lies within the bioequivalence limits. The second from the top example exhibits a much smaller than 100% *T*/*R* value; however, the low variability allows the 90%CI to lie in the 80–125% range. The three cases at the bottom fail to meet the bioequivalence criterion since the 90%CI either lies outside the 80–125% region (the first from the bottom) or the upper end of the 90%CI is higher than the 125% *T*/*R* value (the next two cases from the bottom).

According to the EMA bioequivalence guideline [2] issued in 2010, in specific cases of products with a narrow therapeutic index, the acceptance interval for AUC and C_{max} should be tightened to 90.00–111.11%. The EMA guideline does not define a set of criteria to categorize drugs as narrow therapeutic index drugs; this has to be decided case by case based on clinical considerations.

6.2 Highly Variable Drugs or Drug Products (HVD)

Figure 6.1 includes an example of a highly variable drug or product marked as HVD. Although the T/R estimate is close to unity, both ends of the 90%CI are not within the 80–125% range; this is indicative of the high variability of the drug or the drug product(s). The main problem of HVD is illustrated in Fig. 6.2. This graph shows the

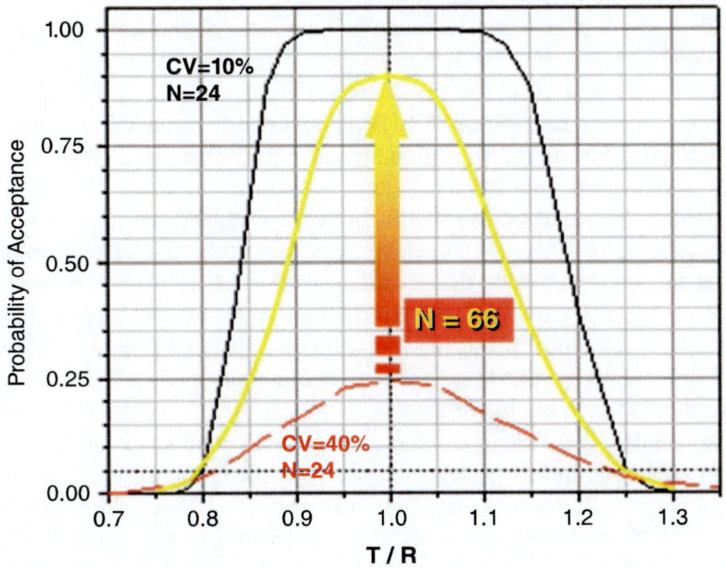

Fig. 6.2 The % of bioequivalence acceptance as a function of the *T*/*R* (%) of the parameter (AUC or C_{max}) for a crossover study with 24 volunteers assuming two different CV values 10 and 40%. The probability of acceptance is increased dramatically when 66 subjects are included in the study with CV = 40%

Table 6.1 The widening of the acceptance interval as a function of the CV of the study as reported in the European Bioequivalence Guideline [2]

Within-subject CV (%)[a]	Lower limit	Upper limit
30	80.00	125.00
35	77.23	129.48
40	74.62	134.02
45	72.15	138.59
≥50	69.84	143.19

[a] $CV(\%) = 100\sqrt{e^{s_{WR}^2} - 1}$

reduction of %BE acceptance (which is proportional to probability of acceptance) as the coefficient of variation (CV) of the study increases. For example, when the (*T*/*R*) value is 100% and the CV is 10%, the %BE acceptance is 100% while the %BE acceptance is only 25% when CV = 40%. Figure 6.2 also shows the increase of the probability of acceptance when the number of subjects is increased from 24 to 66 (notice that the number of subjects N is in the denominator of Eq. (6.1)).

The problem with HVD started in the late 1980s; it was the subject of Bio-International conferences in 1989, 1992, and 1994 [3–5]. Basically, HVD is those drugs that generate an intra-subject coefficient of variation (CV) greater than 30% as measured by the residual CV from the analysis of variance. Under these circumstances, EMA [2] recommends a replicate 3- or 4-period crossover design study. Moreover, bioequivalence in terms of C_{max} can be assessed with a widened acceptance range provided that a wider difference in C_{max} is considered clinically irrelevant based on a sound clinical justification.

The widened acceptance interval must be based on the replicate design where it has been demonstrated that the within-subject variability (or more precisely, the coefficient of variation (CV) of the within-subject variability) for C_{max} of the reference compound in the study is >30%. The reliability of the CV estimate should be proven by the applicant, i.e., a justification is required ruling out that the high CV value is not associated with the presence of outliers in the data. The extent of the widening is defined based on the within-subject variability seen in the bioequivalence study using scaled-average-bioequivalence according to $[U, L] = \exp[\pm k s_{WR}]$, where U is the upper limit of the acceptance range, L is the lower limit of the acceptance range, k is the regulatory constant set to 0.760 and s_{WR} is the within-subject standard deviation of the log-transformed values of C_{max} of the reference product. Table 6.1 gives the U, L limits as a function of the CV of the study as reported in the EMA Guideline [2].

It can be seen that the widening of the acceptance interval levels off for CV values higher than 50%. The interested reader can find in the literature [6, 7] the origin of the bioequivalence limits with leveling off properties. According to the EMA Guideline [2], the widening of the acceptance criteria applies only to C_{max}; for AUC the acceptance range remains at 80.00–125.00% regardless of variability. It should be noted that the publication of the EMA Guideline and the adoption of the bioequivalence limits with leveling off properties diminished the interest in the assessment of bioequivalence of HVD. The methodology introduced in the EMA guideline [2] and the relevant Table 6.1 have been applied extensively. A comparison of EMA and FDA approaches for HVD can be found in the literature [8].

6.3 Epilogue

The exact reason for the use of the term "rate" in the definition of bioavailability is unknown. The use of C_{max} as the sole metric for the rate of drug absorption is probably related to Eq. (1.3); this is so since at t_{max} the rate of input becomes equal to the rate of output in accord with the prevailing hypothesis of Dost of first-order absorption and elimination in 1953. The obvious relationship between C_{max} and the extent of drug absorption (see Eq. 1.3) leads to the conclusion that the use of the "rate" concept is not justified as a valid, independent entity in the definition of bioavailability. Despite these shortcomings, all simulations concerning the rate of oral drug absorption were always routinely based on Eq. (1.3) [3–8].

The term "rate" also has a predominant role in the World Health Organization's (WHO) definition of bioequivalence: "Two pharmaceutical products are bioequivalent if they are pharmaceutically equivalent or pharmaceutical alternatives, and their bioavailabilities, in terms of rate (C_{max} and t_{max}) and extent of absorption (area under the curve), after administration of the same molar dose under the same conditions, are similar to such a degree that their effects can be expected to be essentially the same."

However, we should also recall here the words of Tucker et al. [9]: "*... the ambiguity in the rationale for bioequivalence testing would be removed if the term "rate" was deleted from its definition.*" Their reasoning was based on simulation studies indicating that *"there is no rate parameter which allows products to be compared for both pharmaceutical quality (actual release rate) and clinical safety and efficacy."*

In Chap. 8, we reconsider the rate of absorption concept in the bioequivalence assessment. This reconsideration after so many years coincides with Thales of Miletus words in the subtitle of this chapter, which points to the understanding of the phenomena as time goes by.

References

1. Schuirmann DJ (1987) A comparison of the two one-sided tests procedure and the power approach for assessing the equivalence of average bioavailability. J Pharmacokinet Biopharm 15:657–680. https://doi.org/10.1007/BF01068419
2. EMA (2010) Guideline on the investigation of bioequivalence. London
3. McGilveray IJ, Midha KK, Skelly JP, Dighe S, Doluisio JT, French IW, Karim A, Burford R (1990) Consensus report from "bio international '89": issues in the evaluation of bioavailability data. J Pharm Sci 79:945–946. https://doi.org/10.1002/jps.2600791022

4. Blume HH, Midha KK (1993) Bio-international 92, conference on bioavailability, bioequivalence, and pharmacokinetic studies. J Pharm Sci 82:1186–1189. https://doi.org/10.1002/jps.2600821125
5. McGilveray IJ (1995) An overview of problems and progress at bio-internationals '89 and '92. In: Blume H, Midha K (eds) Bio-international 2, bioavailability, bioequivalence and pharmacokinetic studies. Medpharma Scientific publishers, Stuttgart, pp 109–115
6. Kytariolos J, Karalis V, Macheras P, Symillides M (2006) Novel scaled bioequivalence limits with leveling-off properties. Pharm Res 23:2657–2664. https://doi.org/10.1007/s11095-006-9107-1
7. Karalis V, Symillides M, Macheras P (2011) On the leveling-off properties of the new bioequivalence limits for highly variable drugs of the EMA guideline. Eur J Pharm Sci 44:497–505. https://doi.org/10.1016/j.ejps.2011.09.008
8. Karalis V, Symillides M, Macheras P (2012) Bioequivalence of highly variable drugs: a comparison of the newly proposed regulatory approaches by FDA and EMA. Pharm Res 29:1066–1077. https://doi.org/10.1007/s11095-011-0651-y
9. Tucker GT, Rostami-Hodjegan A, Jackson PR (1995) Bioequivalence-A measure of therapeutic equivalence? In: Blume H, Midha K (eds) Bio-International 2, Bioavailability, bioequivalence and pharmacokinetic studies. Medpharma, Scientific publishers, Stuttgart, pp 35–43

Chapter 7
Bioavailability Under the Prism of Finite Absorption Time

Student: Dr. Einstein, aren't these the same questions as last year's [physics] final exam?
Dr. Einstein: Yes. But this year the answers are different.

Contents

Abstract The introduction of the finite absorption time (FAT) causes a paradigm shift in oral pharmacokinetics. The novel aspects of bioavailability are described in terms of FAT. Explicit relationships are derived for the bioavailable fraction as a function of the model parameters assuming one-compartment model disposition with zero- or first-order absorption lasting for a certain period of time. An application using a theophylline example is presented. Extent and rate metrics are reinterpreted.

P. Macheras, A. A. Tsekouras, *Revising Oral Pharmacokinetics, Bioavailability and Bioequivalence Based on the Finite Absorption Time Concept*,
https://doi.org/10.1007/978-3-031-20025-0_7

Keywords Bioavailability · Oral drug absorption · Finite absorption time · Absorption metrics

7.1 The Columbus' Egg: Drug Absorption Takes Place Under Sink Conditions for Finite Time

The finite absorption time (FAT) concept introduced in Chap. 3 combined with the sink conditions prevailing during drug absorption [1] causes a paradigm shift in oral drug absorption. The real reason for the paradigm shift is shown diagrammatically in a schematic for the underlying processes in the gastrointestinal (GI) membrane/vena cava (V.C.) region, Fig. 7.1.

We consider Fig. 7.1 as a "Columbus egg" [2] since the underlying microscopic processes were not known at the beginning of pharmacokinetics [3], but they have been very well known for several decades now. However, it was only recently realized that the high blood flow (20–40 cm/s) in vena cava ensures sink conditions for the drug transfer [4–6]. In fact, this blood flow rate is five orders of magnitude higher than the usual drug effective permeability estimates ~10^{-4} cm/s. Hence, the rate of presentation of drug to the liver is the product of this blood flow and the drug's concentration in blood which changes linearly in accord with its permeability expressed in velocity units (cm/s), Fig. 7.1. Plausibly, this constant drug input entry to the liver terminates, when either the drug has been completely absorbed prior to its passage from the absorptive sites in the intestines or the dissolved and undissolved drug species pass beyond the absorptive sites; the latter, in the great majority of cases, are located in the small intestines. It should be noted that permeability estimates have been measured for a large number of drugs since permeability is one of the two properties (together with solubility) used for biopharmaceutical classification purposes (see Fig. 2.3) in the relevant FDA [7] and EMA [8] guidelines. For example, due to its permeability, metoprolol is

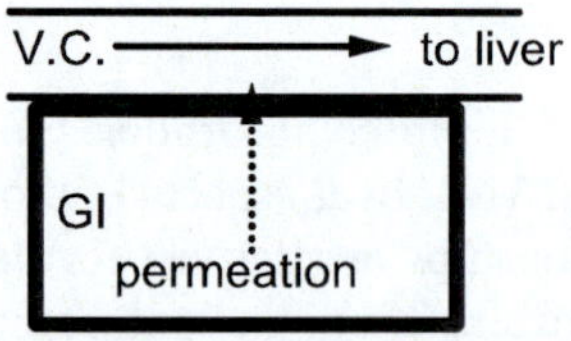

Fig. 7.1 The passive transport of drug molecules (vertical arrow) from the GI tract to blood in vena cava (V.C.) always takes place under sink conditions, since the blood flow rate is very high, 20–40 cm/s [1] (horizontal arrow), resulting in constant drug input rate to the liver

widely reported in the literature as a high-permeability model compound and used as such by FDA.

All these advances justify the abstract notion that Physiologically Based Finite Time Pharmacokinetic (PBFTPK) models are the end of the beginning in oral pharmacokinetics. In other words, the era of oral drug absorption governed by the false concept of first-order absorption and the associated first-order absorption rate constant has been terminated; a new era emerges, which relies on the physiologically sound concept of finite absorption time (FAT). The following section focuses on the fundamental pharmacokinetic changes associated with the replacement of the absorption rate constant with the FAT.

7.2 A Paradigm Shift in Oral Drug Absorption

A logical consequence of the FAT concept is the resulting variations of drug concentration in the blood.

The conceptual difference between Fig. 7.2a–c lies in the fact that beyond time τ for the latter two C, t profiles only elimination takes place since absorption has ceased at time τ. This characteristic has obvious consequences for the assessment of the drug's bioavailability as delineated in the next sections.

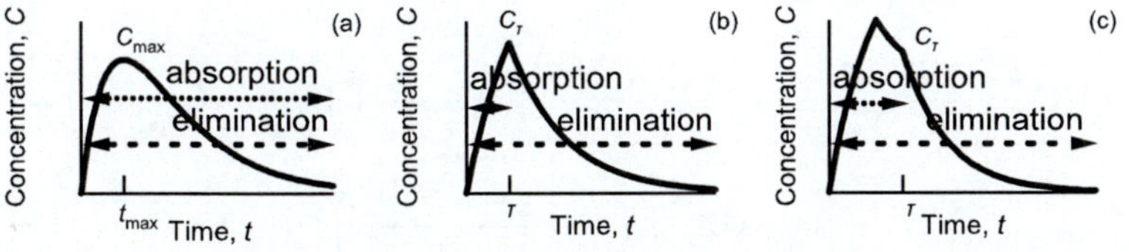

Fig. 7.2 A paradigm shift in oral drug absorption. (**a**) According to the established view, drug absorption and elimination operate concurrently from zero time to infinity [3]. (**b**, **c**) According to the FAT concept [4–6] developed in Chaps. 3 and 4, drug absorption and elimination operate concurrently from zero to τ, while only elimination continues to operate until infinity. Two different profiles can be observed with (**b**) $t_{max} = \tau$ and (**c**) $t_{max} < \tau$. Such behavior has been observed in a number of drugs [6] including paracetamol, cyclosporine, and in axitinib [9] formulations, respectively

7.3 Bioavailability Parameters Under the Prism of Finite Absorption Time Concept

7.3.1 *One Compartment Model with One Constant Input Rate Operating for Time τ*

For this model [5], the following equation was used to describe the drug blood concentration for $t \leq \tau$ assuming termination of absorption at time τ:

$$C(t) = \frac{FD}{\tau} \frac{1}{V_d k_{el}} \left(1 - e^{-k_{el} t}\right) \quad t \leq \tau \tag{7.1}$$

while for $t > \tau$, Eq. (7.2) applies.

$$C(t) = C(\tau) e^{-k_{el}(t-\tau)} \tag{7.2}$$

The drug blood concentration $C(\tau)$ corresponding to time τ, is derived from Eq. (7.1) using $t = \tau$:

$$C(\tau) = \frac{FD}{\tau} \frac{1}{V_d k_{el}} \left(1 - e^{-k_{el}\tau}\right) = \frac{FD}{\tau} \frac{1}{CL} \left(1 - e^{-k_{el}\tau}\right) \tag{7.3}$$

while the areas $[\mathrm{AUC}]_0^\tau$ and $[\mathrm{AUC}]_\tau^\infty$ are derived by integrating Eqs. (7.1) and (7.2), respectively:

$$[\mathrm{AUC}]_0^\tau = \frac{FD}{\tau} \frac{1}{V_d k_{el}} \left(\tau - \frac{1 - e^{-k_{el}\tau}}{k_{el}}\right) = \frac{FD}{V_d k_{el}} \left(1 - \frac{1 - e^{-m \ln 2}}{m \ln 2}\right) = \\ = [\mathrm{AUC}]_0^\infty \left(1 - \frac{1 - e^{-m \ln 2}}{m \ln 2}\right) \tag{7.4}$$

where m is the ratio ($m = \tau / t_{1/2}$) of τ over the half-life $t_{1/2}$ while $k_{el} = (\ln 2)/t_{1/2}$.

$$[\mathrm{AUC}]_\tau^\infty = \frac{C(\tau)}{k_{el}} = \frac{FD}{V_d k_{el}} \frac{1}{k_{el}\tau} \left(1 - e^{-k_{el}\tau}\right) = [\mathrm{AUC}]_0^\infty \frac{1}{m \ln 2} \left(1 - e^{-m \ln 2}\right) \tag{7.5}$$

The sum of the two last integrals, Eqs. (7.4) and (7.5), gives $[\mathrm{AUC}]_0^\infty$, Eq. (1.4).

A hypothetical curve corresponding to the same dose given as an intravenous bolus dose would follow the same track for $t \geq \tau$. Having the general form of:

$$C_{iv}(t) = G e^{-k_{el} t} \tag{7.6}$$

Requiring

$$C_{iv}(t) = C(t) \quad \text{for } t \geq \tau, \tag{7.7}$$

we get:

$$Ge^{-k_{\mathrm{el}}t} = C(\tau)e^{-k_{\mathrm{el}}(t-\tau)} = \frac{FD}{\tau}\frac{1}{V_{\mathrm{d}}k_{\mathrm{el}}}\left(1 - e^{-k_{\mathrm{el}}\tau}\right)e^{-k_{\mathrm{el}}(t-\tau)} \tag{7.8}$$

giving

$$G = \frac{FD}{\tau}\frac{1}{V_d k_{\mathrm{el}}}\left(1 - e^{-k_{\mathrm{el}}\tau}\right)e^{k_{\mathrm{el}}\tau} = \frac{FD}{\tau}\frac{1}{V_{\mathrm{d}}k_{\mathrm{el}}}\left(e^{k_{\mathrm{el}}\tau} - 1\right) \tag{7.9}$$

Then, the hypothetical curve would give:

$$[\mathrm{AUC}_{iv}]_0^\infty = \frac{G}{k_{\mathrm{el}}} = \frac{FD}{V_d k_{\mathrm{el}}}\frac{1}{k_{\mathrm{el}}\tau}\left(e^{k_{\mathrm{el}}\tau} - 1\right) = [\mathrm{AUC}]_0^\infty \frac{1}{k_{\mathrm{el}}\tau}\left(e^{k_{\mathrm{el}}\tau} - 1\right) \tag{7.10}$$

Rearranging the last equation, we get:

$$F = \frac{[\mathrm{AUC}]_0^\infty}{[\mathrm{AUC}_{iv}]_0^\infty} = \frac{k_{\mathrm{el}}\tau}{e^{k_{\mathrm{el}}\tau} - 1} \tag{7.11}$$

where F is the fraction of the dose absorbed since both oral and hypothetical intravenous data rely on a single oral administration of the drug dose to an individual. However, if first-pass effect is not encountered, then F in Eq. (7.11) denotes the bioavailable fraction.

7.3.2 *One Compartment Model with More than One Constant Input Rates Operating for a Total Time* τ

When more than one constant input rate operate successively under in vivo conditions, the concepts developed above can be adapted accordingly. Assuming two constant drug input rates, operating successively, Eqs. (7.1) and (7.2) are replaced by:

$$C(t) = \frac{F_1 D}{\tau_1}\frac{1}{V_{\mathrm{d}}k_{\mathrm{el}}}\left(1 - e^{-k_{\mathrm{el}}t}\right) \qquad 0 < t \leq \tau_1 \tag{7.12}$$

$$C(t) = C(\tau_1)e^{-k_{el}(t-\tau_1)} + \frac{F_2 D}{\tau_2}\frac{1}{V_d k_{el}}\left(1 - e^{-k_{el}(t-\tau_1)}\right) \quad \tau_1 \le t \le \tau_2 \tag{7.13}$$

$$C(t) = C(\tau_1 + \tau_2)e^{-k_{el}(t-\tau_1-\tau_2)} \quad \tau_1 + \tau_2 < t \tag{7.14}$$

If we substitute the values at the turning points via the following expressions:

$$C(\tau_1) = \frac{F_1 D}{\tau_1 V_d k_{el}}\left(1 - e^{-k_{el}\tau_1}\right) \tag{7.15}$$

$$C(\tau_1 + \tau_2) = C(\tau_1)e^{-k_{el}(\tau_1+\tau_2-\tau_1)} + \frac{F_2 D}{\tau_2 V_d k_{el}}\left(1 - e^{-k_{el}(\tau_1+\tau_2-\tau_1)}\right) = \tag{7.16}$$

$$= \frac{F_1 D}{\tau_1 V_d k_{el}}\left(1 - e^{-k_{el}\tau_1}\right)e^{-k_{el}\tau_2} + \frac{F_2 D}{\tau_2 V_d k_{el}}\left(1 - e^{-k_{el}\tau_2}\right) \tag{7.17}$$

we get the final forms for Eqs. (7.13) and (7.14):

$$C(t) = \frac{F_1 D}{\tau_1 V_d k_{el}}\left(1 - e^{-k_{el}\tau_1}\right)e^{-k_{el}(t-\tau_1)} + \frac{F_2 D}{\tau_2 V_d k_{el}}\left(1 - e^{-k_{el}(t-\tau_1)}\right) \tag{7.18}$$

$$C(t) = \left[\frac{F_1 D}{\tau_1 V_d k_{el}}\left(1 - e^{-k_{el}\tau_1}\right)e^{-k_{el}\tau_2} + \frac{F_2 D}{\tau_2 V_d k_{el}}\left(1 - e^{-k_{el}\tau_2}\right)\right]e^{-k_{el}(t-\tau_1-\tau_2)} \tag{7.19}$$

If three drug constant input rates operate successively under in vivo conditions, Eqs. (7.1) and (7.2) are replaced by:

$$C(t) = \frac{F_1 D}{\tau_1}\frac{1}{V_d k_{el}}\left(1 - e^{-k_{el}t}\right) \quad 0 < t \le \tau_1 \tag{7.20}$$

$$C(t) = C(\tau_1)e^{-k_{el}(t-\tau_1)} + \frac{F_2 D}{\tau_2}\frac{1}{V_d k_{el}}\left(1 - e^{-k_{el}(t-\tau_1)}\right) \quad \tau_1 < t \le \tau_1 + \tau_2 \tag{7.21}$$

$$C(t) = C(\tau_1 + \tau_2)e^{-k_{el}(t-\tau_1-\tau_2)} + \frac{F_3 D}{\tau_3}\frac{1}{V_d k_{el}}\left(1 - e^{-k_{el}(t-\tau_1-\tau_2)}\right) \quad \tau_1 + \tau_2 < t \le \tau_1 + \tau_2 + \tau_3 \tag{7.22}$$

$$C(t) = C(\tau_1 + \tau_2 + \tau_3)e^{-k_{el}(t-\tau_1-\tau_2-\tau_3)} \quad \tau_1 + \tau_2 + \tau_3 < t \tag{7.23}$$

or, after substituting with the following expressions for Eqs. (7.21)–(7.23)

$$C(\tau_1) = \frac{F_1 D}{\tau_1 V_d k_{el}}\left(1 - e^{-k_{el}\tau_1}\right) \tag{7.24}$$

$$C(\tau_1 + \tau_2) = \frac{F_1 D}{\tau_1 V_d k_{el}}\left(1 - e^{-k_{el}\tau_1}\right)e^{-k_{el}\tau_2} + \frac{F_2 D}{\tau_2 V_d k_{el}}\left(1 - e^{-k_{el}\tau_2}\right) \tag{7.25}$$

$$C(\tau_1+\tau_2+\tau_3)=\frac{F_1D}{\tau_1V_dk_{el}}\left(1-e^{-k_{el}\tau_1}\right)e^{-k_{el}(\tau_2+\tau_3)}+\frac{F_2D}{\tau_2V_dk_{el}}\times\left(1-e^{-k_{el}\tau_2}\right)e^{-k_{el}\tau_3}+\frac{F_3D}{\tau_3V_dk_{el}}\left(1-e^{-k_{el}\tau_3}\right) \tag{7.26}$$

$$C(t)=\frac{F_1D}{\tau_1V_dk_{el}}\left(1-e^{-k_{el}\tau_1}\right)e^{-k_{el}(t-\tau_1)}+\frac{F_2D}{\tau_2}\frac{1}{V_dk_{el}}\times\left(1-e^{-k_{el}(t-\tau_1)}\right)\quad \tau_1<t\leq\tau_1+\tau_2 \tag{7.27}$$

$$C(t)=\left[\frac{F_1D}{\tau_1V_dk_{el}}\left(1-e^{-k_{el}\tau_1}\right)e^{-k_{el}\tau_2}+\frac{F_2D}{\tau_2V_dk_{el}}\left(1-e^{-k_{el}\tau_2}\right)\right]e^{-k_{el}(t-\tau_1-\tau_2)}+\frac{F_3D}{\tau_3}\frac{1}{V_dk_{el}}\left(1-e^{-k_{el}(t-\tau_1-\tau_2)}\right)\quad \tau_1+\tau_2<t\leq\tau_1+\tau_2+\tau_3 \tag{7.28}$$

$$C(t)=\left[\frac{F_1D}{\tau_1V_dk_{el}}\left(1-e^{-k_{el}\tau_1}\right)e^{-k_{el}(\tau_2+\tau_3)}+\frac{F_2D}{\tau_2V_dk_{el}}\left(1-e^{-k_{el}\tau_2}\right)e^{-k_{el}\tau_3}+\frac{F_3D}{\tau_3V_dk_{el}}\left(1-e^{-k_{el}\tau_3}\right)\right]e^{-k_{el}(t-\tau_1-\tau_2-\tau_3)}\quad \tau_1+\tau_2+\tau_3<t \tag{7.29}$$

7.3.3 One Compartment Model with First-Order Absorption Lasting for Time τ and First-Order Elimination

This model relies on the classical Bateman Eq. (1.2) operating for $t\leq\tau$ and Eq. (7.2) for $t>\tau$.

Using

$$C(\tau)=\frac{FDk_a}{V_d(k_a-k_{el})}\left(e^{-k_{el}\tau}-e^{-k_a\tau}\right) \tag{7.30}$$

the areas $[\mathrm{AUC}]_0^\tau$ and $[\mathrm{AUC}]_\tau^\infty$ are derived by integrating Eqs. (1.2) and (7.2), respectively:

$$[\mathrm{AUC}]_0^\tau=\frac{FD}{V_dk_{el}}-\frac{FDk_a}{V_d(k_a-k_{el})}\left(\frac{e^{-k_{el}\tau}}{k_{el}}-\frac{e^{-k_a\tau}}{k_a}\right) \tag{7.31}$$

$$[\mathrm{AUC}]_\tau^\infty=\frac{C(\tau)}{k_{el}}=\frac{FDk_a}{V_dk_{el}(k_a-k_{el})}\left(e^{-k_{el}\tau}-e^{-k_a\tau}\right) \tag{7.32}$$

The sum of Eqs. (7.31) and (7.32) gives:

$$[\mathrm{AUC}]_0^\infty = \frac{FD}{V_\mathrm{d} k_\mathrm{el}}\left(1 - e^{-k_\mathrm{a}\tau}\right) \tag{7.33}$$

The deviation of the latter quantity from the required value of $\frac{FD}{V_\mathrm{d} k_\mathrm{el}}$ is associated with the discrepancy between the physical assumption that the drug is not absorbed anymore beyond time τ, and the mathematics of first-order absorption process, which lasts until infinity. In fact, the term in parentheses of Eq. (7.33) is linked with the absorption characteristics, i.e., the absorption rate constant k_a and the duration of absorption τ. The impact of this term becomes smaller for high values of k_a and τ, Fig. 7.3; the term used in the ordinate of Fig. 7.3 allows a dimensionless plot.

A hypothetical curve corresponding to the same dose given as an intravenous bolus dose would follow the same track for $t \geq \tau$. Having the general form of:

$$C_{iv}(t) = Ge^{-k_\mathrm{el} t} \tag{7.34}$$

Requiring

$$C_{iv}(t) = C(t) \quad \text{for} \quad t \geq \tau, \tag{7.35}$$

we get:

$$Ge^{-k_\mathrm{el} t} = C(\tau)e^{-k_\mathrm{el}(t-\tau)} = \frac{FDk_\mathrm{a}}{V_\mathrm{d}(k_\mathrm{a} - k_\mathrm{el})}\left(e^{-k_\mathrm{el}\tau} - e^{-k_\mathrm{a}\tau}\right)e^{-k_\mathrm{el}(t-\tau)} \tag{7.36}$$

giving

$$G = \frac{FDk_\mathrm{a}}{V_\mathrm{d}(k_\mathrm{a} - k_\mathrm{el})}\left(e^{-k_\mathrm{el}\tau} - e^{-k_\mathrm{a}\tau}\right)e^{k_\mathrm{el}\tau} = \frac{FDk_\mathrm{a}}{V_\mathrm{d}(k_\mathrm{a} - k_\mathrm{el})}\left(1 - e^{-(k_\mathrm{a} - k_\mathrm{el})\tau}\right) \tag{7.37}$$

Then, the curve for the hypothetical intravenous bolus administration of an equal dose would give:

$$[\mathrm{AUC}_{iv}]_0^\infty = \frac{G}{k_\mathrm{el}} = \frac{FDk_\mathrm{a}}{V_\mathrm{d} k_\mathrm{el}(k_\mathrm{a} - k_\mathrm{el})}\left(1 - e^{-(k_\mathrm{a} - k_\mathrm{el})\tau}\right) \tag{7.38}$$

Using Eqs. (7.33) and (7.38) one can find

$$F = \frac{[\mathrm{AUC}]_0^\infty}{[\mathrm{AUC}_{iv}]_0^\infty} = \left(1 - \frac{k_\mathrm{el}}{k_\mathrm{a}}\right)\frac{1 - e^{-k_\mathrm{a}\tau}}{1 - e^{-(k_\mathrm{a} - k_\mathrm{el})\tau}} \tag{7.39}$$

where F is the fraction of dose absorbed since both oral and intravenous data rely on a single oral dose administration to an individual. However, if first-pass effect is not encountered then F in Eq. (7.39) denotes the bioavailable fraction. The limit of

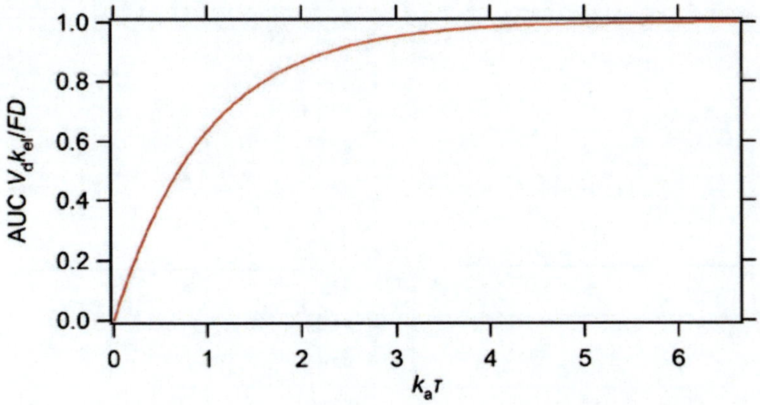

Fig. 7.3 Plot of $[\text{AUC}]_0^{\infty} k_{\text{el}} V_{\text{d}}/FD$ as a function of k_a and τ, (see Eq. 7.33)

Eq. (7.39) for $\tau = 0$ (intravenous bolus dose) correctly predicts $F = 1$ as k_{a} tends to a very high value. Hence, an estimate for F can be obtained, based on the estimates of these parameters derived from the experimental data of oral administration.

A word of caution is required here. The reason for analyzing this model *(One*-compartment *model with first-order absorption lasting for time* τ *and first-order elimination),* is its relevance with the classical Bateman equation, Eq. (1.2). It has been used sporadically in the past [10–12] in experimental and simulation studies in connection with the potential use of truncated concentration-time curves for the assessment of bioequivalence. However, it was shown in Chaps. 3 and 4 that the PBFTPK models based on the successive drug input rates lasting for a specific time period are the most akin to the in vivo conditions and capture both typical and complex drug absorption kinetics phenomena. On the contrary, the Bateman equation relies on the unphysical assumption [3] of infinite absorption time and the time discontinuation is only applied for simulation purposes; it is not based on physical-physiological reasons.

7.4 Extent (Exposure) and Rate Metrics

We consider simple cases for the rate of drug input, i.e., drugs (formulations) exhibiting zero- or first-order input kinetics lasting for time τ, Figs. 7.4 and 7.5.

Zero-order input kinetics. The first two plots (Fig. 7.4a, b) show a simple zero-order process (Fig. 7.4a) and two successive constant input rates (Fig. 7.4b), whereas the termination of the absorption process, τ, leads to $C_{\text{max}} = C(\tau)$. For these cases, (Fig. 7.4a, b), the equality $C_{\text{max}} = C(\tau)$ means that C_{max} is not a steady-state value described by Eq. (1.5), i.e., C_{max} corresponds to the termination of drug input. In some cases, when a very highly soluble and permeable drug is studied, this type of C, t profiles

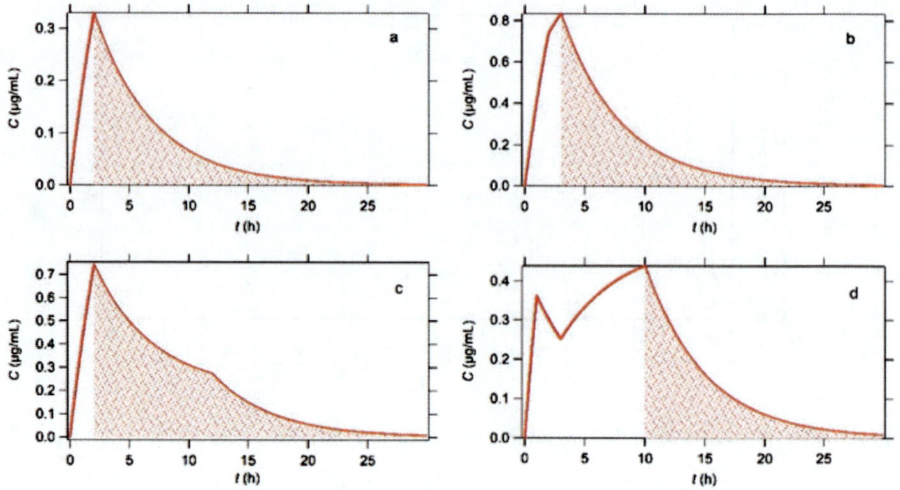

Fig. 7.4 Concentration versus time curves for PBFTPK models assuming different input functions based on zero-order kinetics. In all cases, the value for k_{el} was set equal to 0.2 h^{-1} and $V_d = 100$ L. The absorption/elimination phase of curves in a, b, c, d, was generated using Eqs. (7.1) and (7.2) and its variants Eqs. (7.13)–(7.19). (**a**) $(FD/\tau) = 0.2$ mg/h, $\tau = 2$ h; (**b**) $(F_1D/\tau_1) = 45$ mg/h, $\tau_1 = 2$ h and $(F_2D/\tau_2) = 25$ mg/h, $\tau_2 = 1$ h; (**c**) $(F_1D/\tau_1) = 45$ mg/h, $\tau_1 = 2$ h and $(F_2D/\tau_2) = 4$ mg/h, $\tau_2 = 10$ h; (**d**) $(F_1D/\tau_1) = 40$ mg/h, $\tau_1 = 1$ h and $(F_2D/\tau_2) = 0.5$ mg/h, $\tau_2 = 2$ h and $(F_3D/\tau_3) = 10$ mg/h, $\tau_3 = 7$ h. In all cases, the red areas correspond to drug's elimination phase

(Fig. 7.4a, b) can indicate the completion and not simply the termination of drug absorption; plausibly, biowaivers can exhibit C, t profiles similar to Fig. 7.4a, b.

Figure 7.4c, d show two examples with two or three successive constant input rates, respectively. In both cases, the termination of drug absorption takes place in the colon ($\tau = 12$ h and $\tau = 10$ h), respectively which are longer than t_{max}. In Fig. 7.4c, C_{max} is higher than $C(\tau)$, since the termination of drug absorption lies in the descending portion of the elimination limb of the curve, $C_{max} > C(\tau)$. Figure 7.4d shows a simulated example with three constant input rates causing fluctuation of the drug concentration during the absorption phase. The second lower input rate can be associated with a lower segmental permeability and/or partial drug precipitation. Thus, the observed concentration maximum C_{max} is equal to $C(\tau)$ associated probably with the drug's redissolution.

The simulated examples of Fig. 7.4 demonstrate the rich dynamic behaviors associated with the models with more than one zero-order input rate. Most importantly, Fig. 7.4 shows that the relative magnitude of the parameters C_{max}, t_{max} vis-a-vis $C(\tau)$ and τ can vary remarkably according to the specific case examined. In all cases, however, the concentration of drug starts to decline monotonically beyond the datum point ($C(\tau)$, τ), i.e., drug absorption is not taking place beyond time τ.

First-order input kinetics. Figure 7.5 shows simulated curves based on first-order kinetics lasting for time τ. Three examples with various finite time absorption durations deviating from the classical first-order absorption (top curve in all graphs

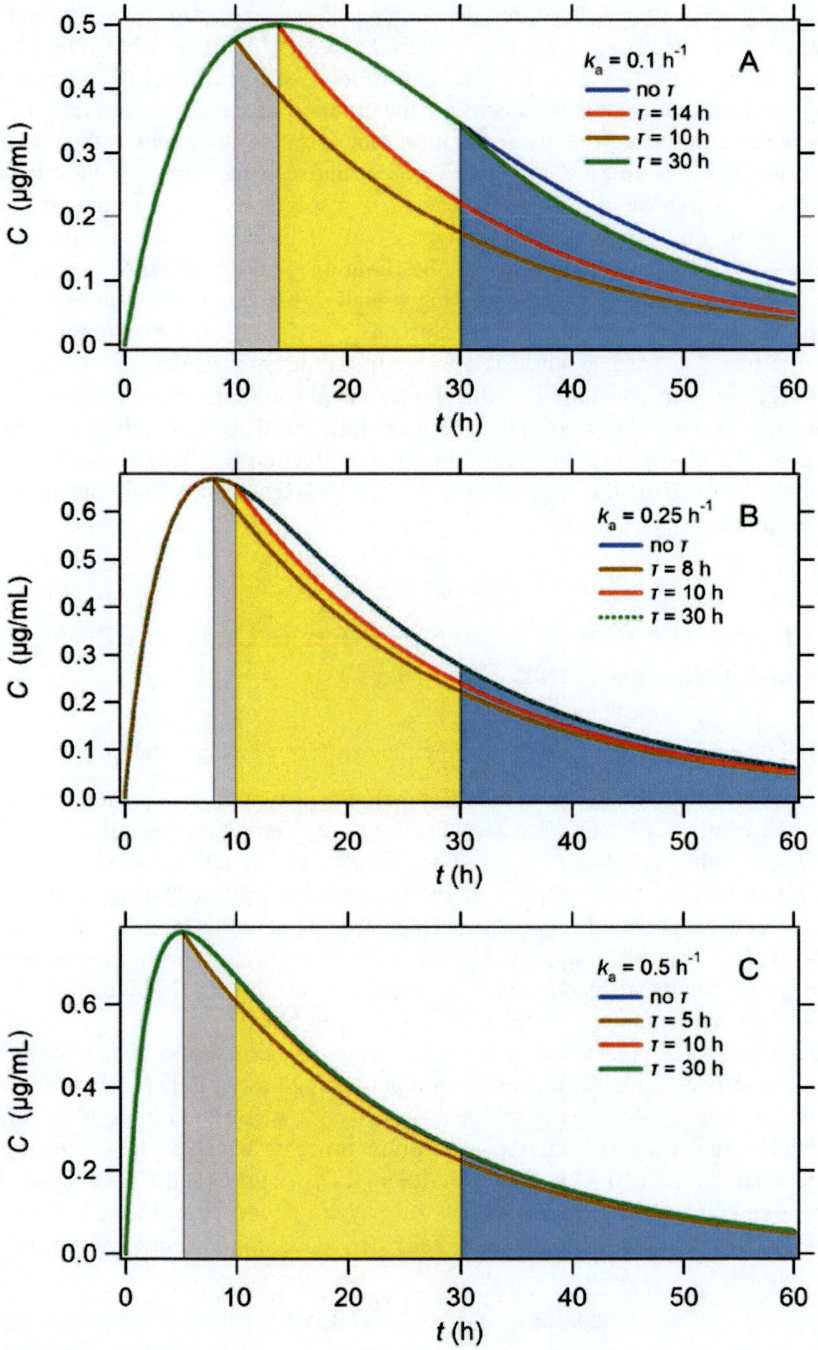

Fig. 7.5 Truncated Bateman (Eq. 1.2) drug concentration profiles with (**a**) $k_a = 0.1\ h^{-1}$, $k_{el} = 0.05\ h^{-1}$ and termination times 10 h (gray), 14 h (yellow) and 30 h (blue); (**b**) $k_a = 0.25\ h^{-1}$, $k_{el} = 0.05\ h^{-1}$ and termination times 8 h (gray), 10 h (yellow) and 30 h (blue); (**c**) $k_a = 0.5\ h^{-1}$, $k_{el} = 0.05\ h^{-1}$, and termination times 5 h (gray), 10 h (yellow) and 30 h (blue)

of Fig. 7.5) are shown using three different values of absorption rate constant k_a, namely, 0.1 h^{-1} (Fig. 7.5a), 0.25 h^{-1} (Fig. 7.5b), 0.5 h^{-1} (Fig. 7.5c). The curves corresponding to the lower value of the absorption rate constant 0.1 h^{-1} depicted in Fig. 7.5a clearly indicate that the smaller the duration of the absorption time is, the larger is the difference in the concentration–time profiles compared to the classical top curve. The examples shown in Fig. 7.5b and c using higher values for the absorption rate constant, 0.25 and 0.5 h^{-1}, respectively, demonstrate that the concentration–time profiles become progressively indistinguishable from the classical case (top curve) as the values of the duration of drug absorption, τ and the absorption rate constant k_a are increasing. Finally, the observations quoted above, which are associated with the relationship $C_{max} \geq C(\tau)$ for the zero-order models shown in Fig. 7.4, are also applicable to the first-order models depicted in Fig. 7.5. Intuitively, one can conclude that the shorter the absorption time duration τ is, the higher is the resemblance of the concentration–time profiles generated from the zero- and first-order models and the classical Bateman function (Eq. 1.2). This is so since all curves approximate the limiting case, i.e., the intravenous bolus administration in the one-compartment model.

7.5 Extent and Rate of Absorption Metrics Under the Prism of Finite Absorption Time (FAT)

We first examine the rate metrics and then the exposure (extent) metrics.

Rate metrics: (C_{max}, t_{max}) vis-a-vis ($C(\tau)$, τ) The use of C_{max} as a measure of the rate of absorption is historically associated with its derivation from Eq. (1.2) as a steady-state value. Although it is used as a bioavailability rate parameter, Eq. (1.5) reveals that C_{max} is also dependent on the extent of absorption. During the previous decades' concerns about this problem were raised and several alternative metrics and methodologies have been suggested [13–17]. However, C_{max} is always being used as a rate parameter in all bioequivalence guidelines, but mainly its numerical value provides the maximum concentration of the drug in blood.

According to Eq. (7.1), which gives the drug concentration at time τ, assuming one-compartment model with zero-order input lasting time τ, $C(\tau)$ is proportional to the rate of input FD/τ. This is an ideal property for a rate metric; besides, time τ underlines the termination of the absorption process, which is the fundamental characteristic of the PBFTPK models. Although C_{max} and $C(\tau)$ differ conceptually, in actual practice the two quantities may or may not be identical since $C_{max} \geq C(\tau)$, Fig. 7.2. When $C_{max} = C(\tau)$, one can easily derive assuming a zero-order input:

$$\text{Rate in} = \frac{V_d C}{dt} = \frac{FD}{\tau} - k_{el} C V_d = 0 \tag{7.40}$$

$$C(\tau) = C_{\max} = \frac{FD}{\tau k_{\mathrm{el}} V_{\mathrm{d}}} = \frac{FD}{\tau \mathrm{CL}} \tag{7.41}$$

This equality means that drug absorption has been terminated or completed at time τ while $C_{\max}$ or $C(\tau)$ is proportional to the input rate (FD/τ) as well as to the extent of absorption (FD), Eq. (7.41). However, $C_{\max}$ or $C(\tau)$ is not the asymptotic limit of a zero-order absorption process with first-order elimination usually found as a steady-state solution in continuous intravenous infusion [18]. In other words, the ($C(\tau)$, τ) datum point is a discontinuity point associated with (1) the completion of the input process (no more drug is available for absorption) or (2) a sudden change in drug's solubility, e.g., precipitation or (3) drug's permeability change, e.g., reduced regional permeability because of pH changes or (4) drug's transit beyond the absorptive sites.

The termination of absorption at time τ for the models with first-order absorption (Fig. 7.5) may result from the passage of the drug beyond the absorptive sites. The corresponding value of $C(\tau)$, is always equal to or smaller than the experimental $C_{\max}$, Fig. 7.2. However, the experimental values for $C(\tau)$ and τ of first-order models are not steady-state values, namely, $C_{\max}$ (Eq. 1.5) and $t_{\max}$ (Eq. 1.6), respectively; the pair ($C(\tau)$, τ) represents a discontinuity time point.

Exposure (extent) metrics: $[\mathrm{AUC}]_0^\infty$ **versus** $[\mathrm{AUC}]_0^\tau$ **and** $[\mathrm{AUC}]_\tau^\infty$ The golden standard for the extent of absorption in bioavailability–bioequivalence studies is $[\mathrm{AUC}]_0^\infty$, Eq. (1.4). This is also justified here mathematically since the sum of Eqs. (7.4) and (7.5), adhering to the model with a single zero-order input, is equal to $[\mathrm{AUC}]_0^\infty$, Eq. (1.4). Although Eq. (7.4) reveals that $[\mathrm{AUC}]_0^\tau$ is a fraction of $[\mathrm{AUC}]_0^\infty$, its magnitude is solely determined from the quantity m, namely, the ratio of duration of the absorption process τ over the elimination half-life, ($m = \tau/t_{1/2}$). Therefore, the meaning of $[\mathrm{AUC}]_0^\tau$ for the zero-order model is not in accord with the usual concept of partial areas used as indicators for the initial rate of exposure [14, 15, 17]. Besides, $[\mathrm{AUC}]_0^\tau$ for the first-order model is dependent on τ (Eq. 7.31), while $[\mathrm{AUC}]_0^\infty$ (Eq. 7.33) is also dependent on τ. Hence, for both zero-and first-order models the usual role of partial areas (portions of $[\mathrm{AUC}]_0^\tau$) is not applicable due to the involvement of τ in the calculations.

According to Eq. (7.5), $[\mathrm{AUC}]_\tau^\infty$ is proportional to the fraction of dose absorbed which remains in the general circulation at time τ [18]. This proportionality is valuable for bioequivalence studies when the duration of the absorption process is short or very short and the absorption phase data exhibit high variability; this is the case with inhalers [19–21] and nasal products [22]. For these formulations, the test-reference comparison can be based on the area $[\mathrm{AUC}]_\tau^\infty$ which is proportional to the fraction of dose absorbed being in the general circulation at time τ. Table 7.1 shows the results based on the analysis of $[\mathrm{AUC}]_\tau^\infty$ for the test and reference formulations of three bioequivalence studies [19–21]. All ratios $\left([\mathrm{AUC}]_\tau^{72}\right)_{\mathrm{test}} / \left([\mathrm{AUC}]_\tau^{72}\right)_{\mathrm{reference}}$ for the five drugs studied lie in the range 0.828–1.104. Although the 90% confidence intervals for the ratio of the means were not constructed, these values lie in the range

Table 7.1 The ratio of $[\mathrm{AUC}]_{\tau}^{72}$ of the test over the reference formulation of five pulmonary drugs in bioequivalence studies

PK parameters	Salmeterol [21]	Fluticasone [21]	Budesonide [20]	Formoterol [20]	Salmeterol [19]
$\frac{([\mathrm{AUC}]_{\tau}^{72})_{\mathrm{test}}}{([\mathrm{AUC}]_{\tau}^{72})_{\mathrm{reference}}}$	1.012	1.104	0.828	0.935	1.074
$\frac{([\mathrm{AUC}]_{\tau}^{72})_{\mathrm{reference}}}{([\mathrm{AUC}]_{0}^{72})_{\mathrm{reference}}}$	0.987	0.858	0.977	0.994	0.935
$\frac{([\mathrm{AUC}]_{\tau}^{72})_{\mathrm{test}}}{([\mathrm{AUC}]_{0}^{72})_{\mathrm{test}}}$	0.986	0.920	0.946	0.985	0.999

Table 7.2 The ratio of $[\mathrm{AUC}]_{\tau}^{14}$ of the test over the reference formulation in a nasal absorption bioequivalence study of budesonide [23] and $[\mathrm{AUC}]_{\tau}^{3}$ or $[\mathrm{AUC}]_{\tau}^{4}$ ratios (powder vs. solution) in a comparative systemic bioavailability study [24] of three nasal remimazolam (RMZ) formulations

PK parameters	Budesonide	PK parameters	RMZ (10 mg)	PK parameters	RMZ (20 mg)	RMZ (40 mg)
$\frac{([\mathrm{AUC}]_{\tau}^{14})_{\mathrm{test}}}{([\mathrm{AUC}]_{\tau}^{14})_{\mathrm{reference}}}$	1.024	$\frac{([\mathrm{AUC}]_{\tau}^{3})_{\mathrm{powder}}}{([\mathrm{AUC}]_{\tau}^{3})_{\mathrm{solution}}}$	1.226	$\frac{([\mathrm{AUC}]_{\tau}^{4})_{\mathrm{powder}}}{([\mathrm{AUC}]_{\tau}^{4})_{\mathrm{solution}}}$	1.422	2.118
$\frac{([\mathrm{AUC}]_{\tau}^{14})_{\mathrm{reference}}}{([\mathrm{AUC}]_{0}^{14})_{\mathrm{reference}}}$	0.942	$\frac{([\mathrm{AUC}]_{\tau}^{3})_{\mathrm{solution}}}{([\mathrm{AUC}]_{0}^{3})_{\mathrm{solution}}}$	0.831	$\frac{([\mathrm{AUC}]_{\tau}^{4})_{\mathrm{solution}}}{([\mathrm{AUC}]_{0}^{4})_{\mathrm{solution}}}$	0.854	0.858
$\frac{([\mathrm{AUC}]_{\tau}^{14})_{\mathrm{test}}}{([\mathrm{AUC}]_{0}^{14})_{\mathrm{test}}}$	0.944	$\frac{([\mathrm{AUC}]_{\tau}^{3})_{\mathrm{powder}}}{([\mathrm{AUC}]_{0}^{3})_{\mathrm{powder}}}$	0.893	$\frac{([\mathrm{AUC}]_{\tau}^{4})_{\mathrm{powder}}}{([\mathrm{AUC}]_{0}^{4})_{\mathrm{powder}}}$	0.858	0.904

80–125% used in bioequivalence testing. Besides, the ratios $[\mathrm{AUC}]_{\tau}^{72}/[\mathrm{AUC}]_{0}^{72}$ for all drugs and formulations studied are in the range 0.858–0.999, Table 7.1, which indicates that the area $[\mathrm{AUC}]_{\tau}^{72}$ represents a very large portion (>80%) of the total area $[\mathrm{AUC}]_{0}^{\infty}$. Since the variability of the experimental data in the ascending limb of the curve of the inhaled products is very high (19–21), while a dense sampling strategy is usually applied, the use of $[AUC]_{\tau}^{72}$ as an extent of absorption metric can lead to a smaller number of volunteers and a less dense sampling protocol in bioequivalence studies. Additional relevant data from two nasal absorption studies were analyzed assuming $t_{\mathrm{max}} = \tau$ and presented in Table 7.2.

7.6 Scientific-Regulatory Implications

In the light of the previous discussion, a reconsideration of the meaning and use of the typical bioequivalence parameters (C_{max}, t_{max}, and partial areas) is required. This is summarized in Table 7.3 along with the meaning and potential use of the novel parameters ($C(\tau)$, τ, $[\mathrm{AUC}]_{\tau}^{\infty}$).

Table 7.3 The meaning of the classical and novel bioequivalence parameters in the light of zero- and first-order models with drug absorption lasting for time τ

Parameters	Remarks
C_{max}, $C(\tau)$	When $t_{max} = \tau$, C_{max} is equal to $C(\tau)$; it corresponds to the blood concentration at the termination or completion of drug absorption at time τ. When $t_{max} < \tau$, then $C_{max} > C(\tau)$; C_{max} does not correspond to the termination or completion of drug absorption at time τ
t_{max}, τ	When $t_{max} = \tau$ the recorded t_{max} corresponds to the termination or completion of drug absorption at time τ. When $t_{max} < \tau$, the numerical value of τ is the physiologically meaningful parameter, since it denotes the duration of the absorption process
Partial areas (portions of $[\mathrm{AUC}]_0^\tau$) [20]	For the zero-order model, the magnitude of the areas (portions of $[\mathrm{AUC}]_0^\tau$) depends exclusively on m, ($m = \tau/t_{1/2}$); therefore, these portions cannot be used as early absorption rate indicators For the first-order model, the magnitude of the areas (portions of $[\mathrm{AUC}]_0^\tau$) and the total area ($[\mathrm{AUC}]_0^\infty$) are both dependent on τ; therefore, these portions are not typical indicators of the early absorption rate
$[\mathrm{AUC}]_\tau^\infty$	Proportional to the fraction of dose absorbed which remains in the body at time τ. It could be used instead of $[\mathrm{AUC}]_0^\infty$ when very fast absorption is encountered

The remarks quoted in Table 7.3 can guide regulatory agencies for potential changes in the assessment of bioequivalence studies. The utilization of the parameters τ and $[\mathrm{AUC}]_\tau^\infty$ as well as the reconsideration of the partial area utility as a rate of exposure metric are the most challenging questions. In addition, the two recommendations of the current bioequivalence guidelines [7, 8], namely, (1) "The sampling schedule should also cover the plasma concentration–time curve long enough to provide a reliable estimate of the extent of exposure which is achieved if $[\mathrm{AUC}]_0^t$ covers at least 80% of $[\mathrm{AUC}]_0^\infty$ " and (2) the specific time limit of 72 h, for the calculation of total AUC, i.e., "AUC truncated at 72 h ($[\mathrm{AUC}]_0^{72}$) may be used as an alternative to $[\mathrm{AUC}]_0^t$ for comparison of extent of exposure as the absorption phase has been covered by 72 h for immediate release formulations," should be reconsidered in view of the finite absorption time (FAT) concept (see Chaps. 3 and 4). In addition, drug absorption beyond 30 h is not physiologically sound [4, 25].

7.7 Toward the Unthinkable: Estimation of Absolute Bioavailability from Oral Data Exclusively

In Sect. 7.3, we derived equations that can be used for the estimation of fraction of dose absorbed or the absolute bioavailable fraction when first-pass effect is not encountered, F from oral data exclusively. In these equations like (7.11) and (7.39), F is expressed as a function of τ, k_{el} and k_a.

Estimation of F for one-compartment model drugs using a proportionality ratio of areas For drugs obeying one-compartment model disposition, the estimation of F from oral data exclusively can be also accomplished using a proportionality ratio of areas under the concentration–time curve. If the input kinetics lasts τ time units, an estimate for F can be derived from an areas proportionality corrected in terms of dose:

$$F = \frac{[(\mathrm{AUC})]_{0\ \mathrm{oral}}^{\infty}\mathrm{Dose}}{\left[(\mathrm{AUC})_0^{\infty}\right]_{hy.i.v}\mathrm{FDose}} \tag{7.42}$$

where $\left[(\mathrm{AUC})_0^{\infty}\right]_{hy.i.v}$ corresponds to the area of the hypothetical intravenous bolus administration of the same dose derived from the back extrapolation of the elimination phase experimental oral data beyond time τ of the oral dose. Its numerical value is calculated from the ratio $e^{(y\ -\ \mathrm{intercept})}/k_{\mathrm{el}}$, where the y-intercept on the ln$C$ axis corresponds to the back extrapolated regression line with slope $-k_{\mathrm{el}}$ of ln$C$, t elimination phase data beyond time τ. The integral $[\mathrm{AUC}]_{0\ \mathrm{oral}}^{\infty}$ is calculated using the trapezoidal rule from the experimental data. Solving Eq. (7.42) in terms of F:

$$F^2 = \frac{[\mathrm{AUC}]_{0\ \mathrm{oral}}^{\infty}}{[\mathrm{AUC}]_{0\ hy.i.v}^{\infty}} \tag{7.43}$$

The positive root of Eq. (7.43) provides the estimate for F.

Analysis of a theophylline bioequivalence study Here, we analyze published data [26] from a bioequivalence study with three formulations of theophylline (Fig. 7.6) to derive estimates for F. This historically first calculation of absolute bioavailability from oral data exclusively [5] is applied to theophylline since its absorption is not problematic being a Class I drug (highly soluble, highly permeable). We first

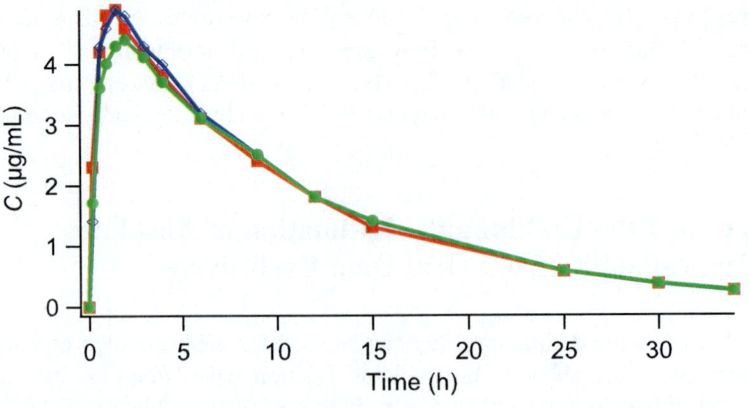

Fig. 7.6 Mean theophylline serum concentration in 18 human subjects who received three different 200 mg theophylline capsules: (▪) Product A; (◇) Product B; and (•) Product C [26]

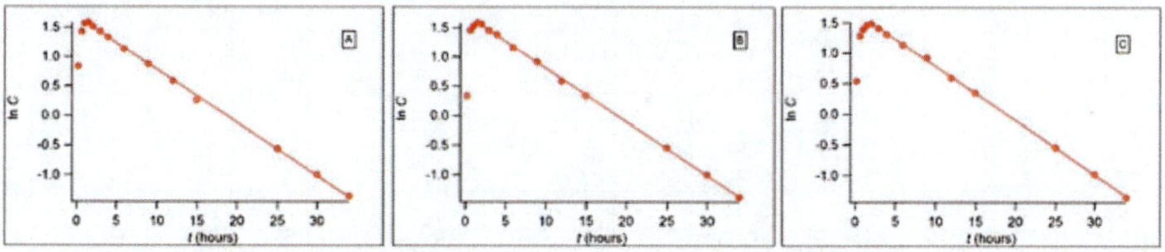

Fig. 7.7 Semi-logarithmic concentration–time plots of theophylline formulations A, B, and C [26]

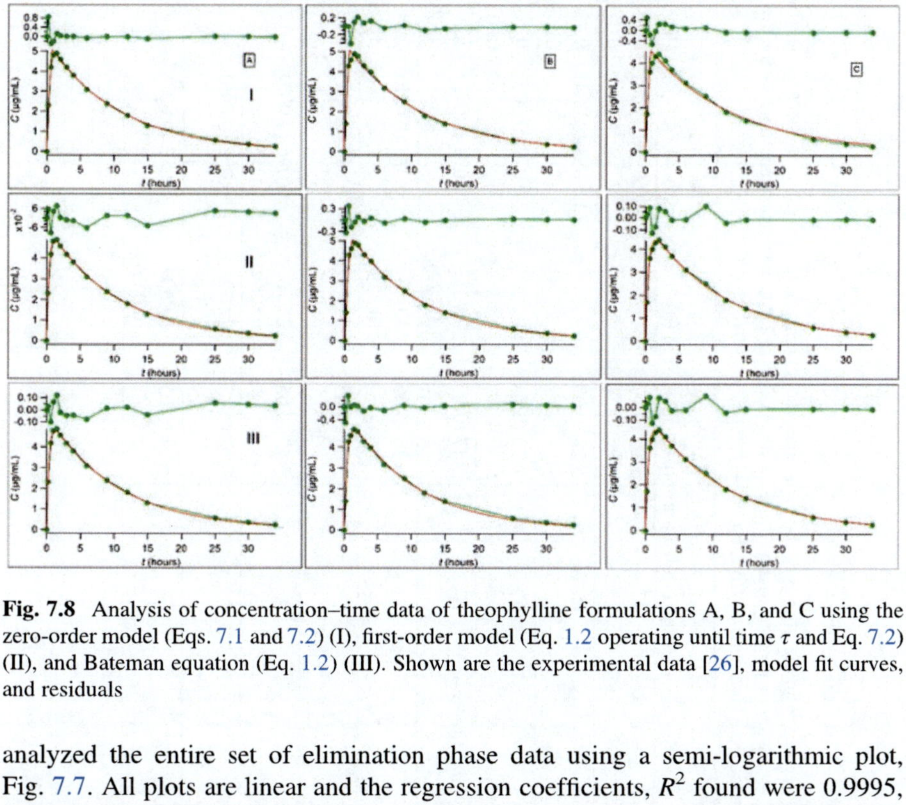

Fig. 7.8 Analysis of concentration–time data of theophylline formulations A, B, and C using the zero-order model (Eqs. 7.1 and 7.2) (I), first-order model (Eq. 1.2 operating until time τ and Eq. 7.2) (II), and Bateman equation (Eq. 1.2) (III). Shown are the experimental data [26], model fit curves, and residuals

analyzed the entire set of elimination phase data using a semi-logarithmic plot, Fig. 7.7. All plots are linear and the regression coefficients, R^2 found were 0.9995, 0.9997, and 0.9998 for formulations A, B, and C, respectively. This verifies that the entire set of elimination phase data follows a one-compartment model disposition. Then, an unrestricted non-linear least squares fit of the zero-order model (Eqs. 7.1 and 7.2), the first-order model (Eq. 1.2 operating until time τ and Eq. 7.2), and Bateman equation (Eq. 1.2 without time restriction) was applied, Fig. 7.8. The parameter estimates are listed in Table 7.4 along with the calculated F values derived from Eqs. (7.11) and (7.39) adhering to the zero- and first-order model, respectively. In addition, estimates for F derived from Eq. (7.43) are listed in Table 7.4; Fig. 7.9 shows the relevant plot for the estimation of $[\mathrm{AUC}]_0^\infty{}_{hy.i.v}$ of formulation A, which is used in Eq. (7.43).

Table 7.4 Parameter estimates (1 σ) derived from the fittings of the zero-order model (Eqs. 7.1 and 7.2), first-order model (Eq. 1.2 operating until time τ and Eq. 7.2), and Bateman equation (Eq. 1.2 without time restriction) to experimental data [26], correlation coefficients for the fits and calculated bioavailable fraction F; the estimates for F designated F_{areas} are derived from Eq. (7.43)

Model	Formulation	FD/V_d (μg/mL)	k_a (h^{-1})	k_{el} (h^{-1})	τ (h)	R^2	F	F_{areas}
Zero-order model (Eqs. 7.1 and 7.2)	A	5.29 (0.17)	–	0.092 (0.008)	0.72 (0.05)	0.990	0.967	0.960
	B	5.28 (0.09)	–	0.088 (0.004)	0.75 (0.03)	0.997	0.967	0.962
	C	4.73 (0.14)	–	0.079 (0.006)	0.76 (0.05)	0.989	0.970	0.959
First-order model (Eq. 1.2 operating until time τ and Eq. 7.2)	A	5.47 (0.09)	2.703 (0.122)	0.094 (0.002)	1.49 (0.30)	0.9997	1.043	–
	B	6.35 (0.66)	1.626 (0.329)	0.093 (0.005)	1.21 (0.18)	0.997	1.036	–
	C	5.03 (0.05)	2.062 (0.085)	0.087 (0.002)	2.93 (3.04)	0.9993	1.451	–
Bateman equation (Eq. 1.2)	A	5.42 (0.05)	2.776 (0.089)	0.096 (0.002)	–	0.9995	–	–
	B	5.63 (0.15)	2.072 (0.180)	0.098 (0.006)	–	0.996	–	–
	C	5.04 (0.05)	2.060 (0.069)	0.087 (0.002)	–	0.9993	–	–

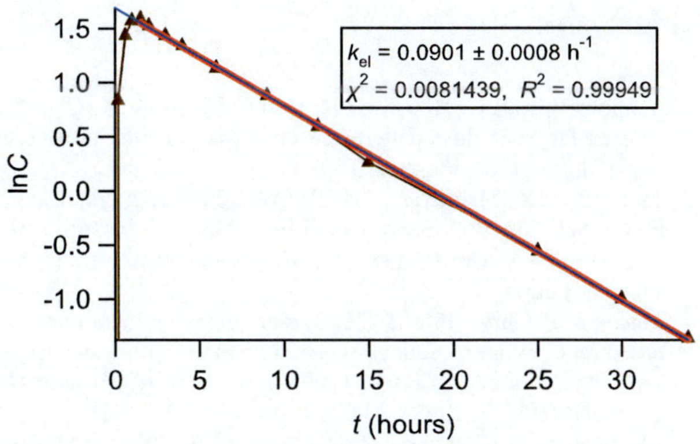

Fig. 7.9 Semi-logarithmic plot of theophylline from formulation A serum data. The "triangle" represents the $[AUC]_0^\infty{}_{hy.i.v}$ semi-logarithmically while the $[AUC]_0^\infty{}_{oral}$ corresponds to the area under the curve of the experimental data points, depicted semi-logarithmically too

Excellent fits were observed for all data sets, Table 7.4. There is a minor superiority of the Bateman function and the first-order model over the zero-order model which is associated with the usually more erratic absorption phase whereas one or two data points deviate slightly from the fitting of the zero-order model. However, Eq. (7.11) provides for F a single estimate, 0.97 for all formulations studied while the estimates for F based on Eq. (7.35) are 1.04 for formulations A and B and 1.45 for formulation C. The latter numerical value originates from the poor estimate for τ, 2.93 (3.04) h derived from the first-order model fitting (Table 7.4). It is very well known that estimates for F cannot be derived from the fitting of the Bateman function to oral data. Nevertheless, all three approaches demonstrate that theophylline absorption has terminated in the small intestine; however, the zero- and first-order model fittings clearly show *the complete absorption of theophylline in the small intestines from the three formulations studied.* Needless to say that no clear advantage of the zero- and first-order model fittings over the classical Bateman equation in terms of the modeling exercise could be concluded. However, the infinite time implied in the use of a first-order input, everyone knows, never happens in the real world. Thus, the use of the finite absorption time limit in the zero- and first-order model fittings allowed the estimation of F. This cannot be accomplished using the classical approach. The estimates for F derived in Table 7.4 are in full agreement with the reported value for F, 0.96 ± 0.03 for immediate release theophylline tablets [27].

The simple example of theophylline analyzed belongs to the case of Class I drugs following a one-compartment model disposition, whereas $C_{max} = C(\tau)$. However, caution should be exercised with the application of this method since most of the drugs exhibit two-compartment model disposition.

References

1. Iranpour P, Lall C, Houshyar R, Helmy M, Yang A, Choi JI, Ward G, Goodwin SC (2016) Altered Doppler flow patterns in cirrhosis patients: an overview. Ultrasonography 35:3–12. https://doi.org/10.14366/usg.15020
2. Tsekouras AA, Macheras P (2022) Columbus' egg: oral drugs are absorbed in finite time. Eur J Pharm Sci 176:106265. https://doi.org/10.1016/j.ejps.2022.106265
3. Dost HF (1953) Der Blutspiegel. Kinetik der Konzentrationsabläufe in der Kreislaufüssigkeit. Thieme, Leipzig
4. Macheras P, Chryssafidis P (2020) Revising pharmacokinetics of oral drug absorption: I Models based on biopharmaceutical/ physiological and finite absorption time concepts. Pharm Res 37: 187. https://doi.org/10.1007/s11095-020-02894-w. (Erratum. Pharm Res 37:206 (2020) https://doi.org/10.1007/s11095-020-02935-4)
5. Chryssafidis P, Tsekouras AA, Macheras P (2022) Revising pharmacokinetics of oral drug absorption: II bioavailability-bioequivalence considerations. Pharm Res 38:1345–1356. https://doi.org/10.1007/s11095-021-03078-w
6. Chryssafidis P, Tsekouras AA, Macheras P (2022) Re-writing oral pharmacokinetics using physiologically based finite time pharmacokinetic (PBFTPK) models. Pharm Res 39:691–701. https://doi.org/10.1007/s11095-022-03230-0
7. Food and Drug Administration (1971) Code of Federal Regulations Title 21, 5 Subpart B - Procedures for Determining the Bioavailability or Bioequivalence of Drug Products. https://www.accessdata.fda.gov/scripts/cdrh/cfdocs/cfcfr/CFRSearch.cfm?fr=320.21
8. European Medicines Agency (2017) Committee for medicinal products for human use (CHMP) guideline on the investigation of bioequivalence. London
9. Alimpertis N, Tsekouras AA, Macheras P (2022) Revising the assessment of bioequivalence in the light of finite absorption time (FAT) concept: The axitinib case. Submitted to 30th PAGE meeting Ljubljana 28 June–1 July, 2022
10. Lovering EG, McGilveray IJ, McMillan I, Tostowaryk W (1975) Comparative bioavailabilities from truncated blood level curves. J Pharm Sci 64:1521–1524. https://doi.org/10.1002/jps.2600640921
11. Sugano K (2012) Biopharmaceutics modeling and simulations: theory, practice, methods, and applications. Wiley
12. Sugano K (2021) Lost in modelling and simulation? ADMET DMPK 9:75–109. https://doi.org/10.5599/admet.923.5
13. Endrenyi L, Fritsch S, Yan W (1991) Cmax/AUC is a clearer measure than Cmax for absorption rates in investigations of bioequivalence. Int J Clin Pharmacol Ther Toxicol 29:394–399
14. Chen ML (1992) An alternative approach for assessment of rate of absorption in bioequivalence studies. Pharm Res 9:1380–1385. https://doi.org/10.1023/a:1015842425553
15. Chen M-L, Davit B, Lionberger R, Wahba Z, Ahn H-Y, Yu LX (2011) Using partial area for evaluation of bioavailability and bioequivalence. Pharm Res 28:1939–1947. https://doi.org/10.1007/s11095-011-0421-x
16. Macheras P, Symillides M, Reppas C (1996) An improved intercept method for the assessment of absorption rate in bioequivalence studies. Pharm Res 13:1755–1758. https://doi.org/10.1023/a:1016421630290
17. Macheras P, Symillides M, Reppas C (1994) The cutoff time point of the partial area method for assessment of rate of absorption in bioequivalence studies. Pharm Res 11:831–834. https://doi.org/10.1023/A:1018921622981
18. Niazi S (1979) Textbook of biopharmaceutics and clinical pharmacokinetics. Appleton-Century-Crofts, New York
19. Soulele K, Macheras P, Silvestro L, Rizea Savu S, Karalis V (2015) Population pharmacokinetics of fluticasone propionate/salmeterol using two different dry powder inhalers. Eur J Pharm Sci 80:33–42. https://doi.org/10.1016/j.ejps.2015.08.009

20. Soulele K, Macheras P, Karalis V (2018) On the pharmacokinetics of two inhaled budesonide/formoterol combinations in asthma patients using modeling approaches. Pulm Pharmacol Ther 48:168–178. https://doi.org/10.1016/j.pupt.2017.12.002
21. Soulele K, Macheras P, Karalis V (2017) Pharmacokinetic analysis of inhaled salmeterol in asthma patients: evidence from two dry powder inhalers. Biopharm Drug Dispos 38:407–419. https://doi.org/10.1002/bdd.2077
22. Bioavailability and bioequivalence studies for nasal aerosols and nasal sprays for local action. FDA Guidance 2003
23. do Carmo Borges NC, Astigarraga RB, Sverdloff CE, Borges BC, Paiva TR, Galvinas PR, Moreno RA (2010) Budesonide quantification by HPLC coupled to atmospheric pressure photoionization (APPI) tandem mass spectrometry. Application to a comparative systemic bioavailability of two budesonide formulations in healthy volunteers. J Chromatography B 879:236–242. https://doi.org/10.1016/j.jchromb.2010.12.003
24. Pesic M, Schippers F, Saunders R, Webster L, Donsbach M, Stoehr T (2020) Pharmacokinetics and pharmacodynamics of intranasal remimazolam-a randomized controlled clinical trial. Eur J Clin Pharmacol 76:1505–1516. https://doi.org/10.1007/s00228-020-02984-z
25. Abuhelwa A, Foster D, Upton R (2016) A quantitative review and meta-models of the variability and factors affecting oral drug absorption-part II: gastrointestinal transit time. AAPS J 18:1322–1333. https://doi.org/10.1208/s12248-016-9953-7
26. Meyer MC, Jarvi EJ, Straughn AB, Pelsor FR, Williams RL, Shah VP (1999) Bioequivalence of immediate-release theophylline capsules. Biopharm Drug Dispos 20:417–419. https://doi.org/10.1002/1099-081X(199912)20:9%3C417::AID-BDD205%3E3.0.CO;2-W
27. Hendeles L, Weinberger M, Bighley L (1977) Absolute bioavailability of oral theophylline. Am J Hosp Pharm 34:525–527. https://doi.org/10.1093/ajhp/34.5.525

Chapter 8
Bioequivalence Under the Prism of Finite Absorption Time

Curiosity has its own reason for existing.
Albert Einstein (1879–1955)

Contents

Abstract According to the finite absorption time, the area under the curve from zero to the end of the absorption processes, τ, namely $[AUC]_0{}^\tau$, is the proper metric for the extent of drug absorption. This concept was successfully applied to two published bioavailability digoxin studies. The cumulative ratio of areas as a function of sampling time is suggested as a diagnostic test and criterion for the assessment of the drug's extent of absorption.

Keywords Bioequivalence · Assessment of bioequivalence · Digoxin · Finite absorption time

Prior to the definition of bioavailability by FDA in 1977 [1], scientists were ambiguous about the proper methodology in terms of the sampling design and the metrics applied for studying bioavailability. One such study was performed by Lovering et al. [2] in 1975. The authors of the study focused on the period of time after administration over which blood level measurements are required to obtain a reliable bioavailability comparison. They analyzed literature data on the following ten drugs: acetaminophen, aminosalicylic acid, chloramphenicol, chlordiazepoxide,

P. Macheras, A. A. Tsekouras, *Revising Oral Pharmacokinetics, Bioavailability and Bioequivalence Based on the Finite Absorption Time Concept*,
https://doi.org/10.1007/978-3-031-20025-0_8

digoxin, isoniazid, phenylbutazone, sulfamethizole, tetracycline, and warfarin. They found that reliable bioavailability comparisons among different brands of the drugs could have been made by using truncated concentration–time curves since the ratios of areas under the curve changed little between the end of the absorption period and the time when blood sampling was terminated. Although the authors [2] assumed first-order absorption kinetics lasting for a specific time period, their findings relied on the calculation of the ratios (test/reference) of the cumulative areas under the blood concentration–time curve as a function of time. Plausibly, this plot levels off upon the completion of the absorption processes of drug from both formulations; we accordingly concluded that this is an explicit piece of evidence that the finite absorption time (FAT) has reached for both formulations. These observations prompted us to re-examine [3] two bioavailability/bioequivalence digoxin studies [4, 5] under the prism of FAT concept.

8.1 Reconsidering Digoxin Bioavailability/Bioequivalence Studies in the Light of Finite Absorption Time Concept

The first study [4] we analyzed [3] compares bioavailability upon administration of a digoxin tablet under fasting or fed conditions. Figure 8.1 shows the concentration–time profile in three subjects. According to the authors [4] "when measured by peak serum digoxin concentration as well as by area under the serum digoxin concentration-time curve, the bioavailability of digoxin appeared to be higher in the fasting state than in the fed state. However, when measured by cumulative five-day urinary excretion of digoxin bioavailability was identical in both conditions," Fig. 8.2.

The termination of digoxin absorption was estimated [3] to be at 1 and 3 h under fasting and fed conditions, respectively. Using the pertinent AUC ratios, i.e.,

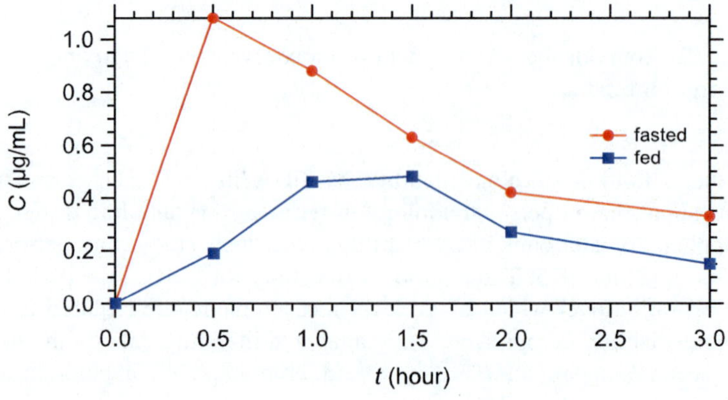

Fig. 8.1 Concentration–time data from a digoxin bioavailability study [4]

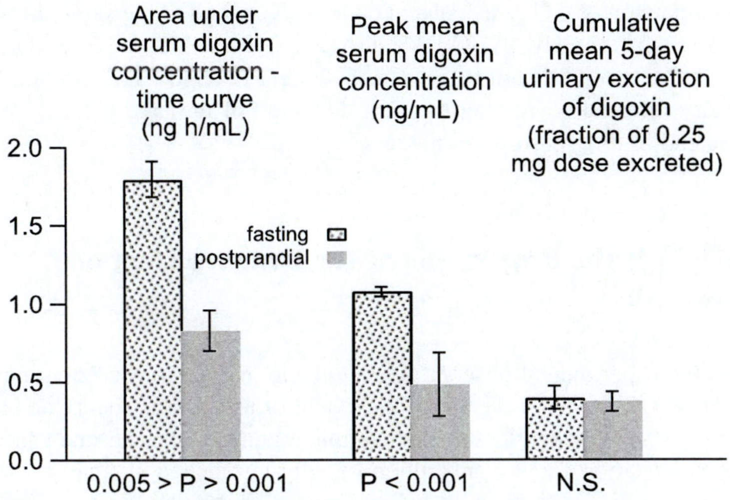

Fig. 8.2 Comparison of bioavailability as estimated by the area under serum concentration, and cumulative urinary excretion of digoxin. N.S. = Not significant [4]

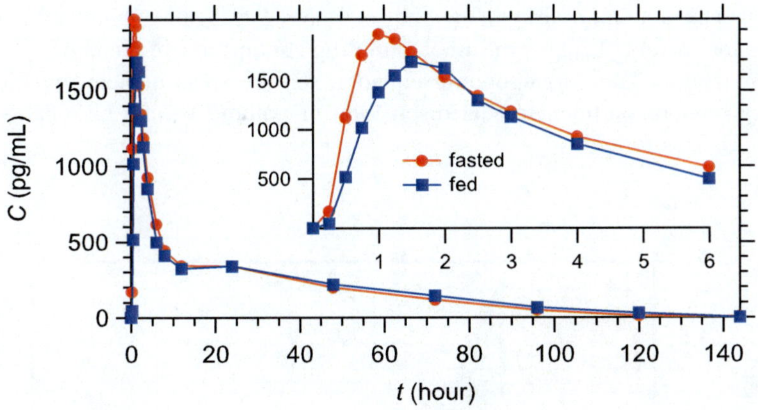

Fig. 8.3 Concentration–time data from a bioequivalence study [5]. Inset shows an expanded view of the first 6 h of the data

$\frac{[\mathrm{AUC}]_0^1 \mathrm{fasted}}{[\mathrm{AUC}]_0^3 \mathrm{fed}}$ we found [3] the same result (equal bioavailability) with the results derived from the cumulative 5 day urinary excretion of digoxin.

The second digoxin study we examined [3] is a 1992 bioequivalence study analyzed by FDA [5], Fig. 8.3. The duration of digoxin absorption in this study under fasting and fed conditions was found to be 1 and 1.5 h, respectively; the corresponding ratios $\frac{[\mathrm{AUC}]_0^1 \mathrm{fasted,test}}{[\mathrm{AUC}]_0^1 \mathrm{fasted,reference}}$, $\frac{[\mathrm{AUC}]_0^{1.5} \mathrm{fed,test}}{[\mathrm{AUC}]_0^1 \mathrm{fed,reference}}$ were quite similar to the classical comparison of AUCs calculated up to the very end of the sampling scheme (144 h) and infinity, namely, $[\mathrm{AUC}]_0^{144}$ and $[\mathrm{AUC}]_0^{\infty}$, reported in the FDA document.

All the above results demonstrate that the extent of digoxin absorption can be equally well estimated by relying on the calculation of the area $[\mathrm{AUC}]_0^{\tau}$, where τ denotes the end of drug absorption process.

8.2 $[\mathrm{AUC}]_0^{\tau}$ Is the Proper Metric for Drug's Extent of Absorption

Apart from the experimental observations quoted in the previous section indicating the utility of $[\mathrm{AUC}]_0^{\tau}$ as an indicator of the extent of absorption, theoretical considerations can be also applied to substantiate this argument. In fact, one can write a mass balance equation for the absorption of drug terminated at time, τ assuming one-compartment model disposition for any type of absorption kinetics, Fig. 8.4:

$$Q_{\tau} = FD - Q_{\mathrm{el}(0-\tau)} \tag{8.1}$$

where Q_{τ} is the amount of drug in the body at time τ, F is the bioavailable fraction of the drug dose D and $Q_{\mathrm{el}(0\text{-}\tau)}$ is the amount of drug eliminated from the body between time zero and τ. The corresponding equation for the areas depicted in Fig. 8.4 assuming one-compartment model disposition of volume V_{d} and elimination rate constant k_{el} is as follows:

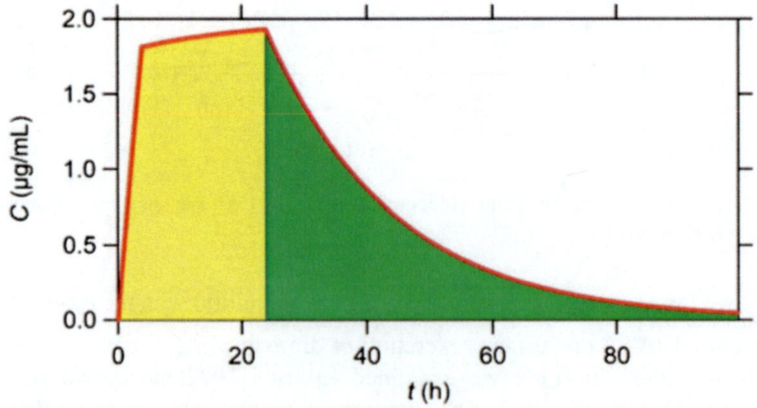

Fig. 8.4 A schematic for a drug absorbed in finite time, τ. The areas $[\mathrm{AUC}]_0^{\tau}$ (on the left, yellow area) and $[\mathrm{AUC}]_{\tau}^{\infty}$ (on the right, green area) are depicted. The C, t profile was generated based on a drug following one-compartment model disposition and two successive constant input rates. The following parameter values were used in the simulation: $F_1D/V_{\mathrm{d}} = 2$ μg/mL, $F_2D/V_{\mathrm{d}} = 2$ μg/mL, $\tau_1 = 4$ h, $\tau_2 = 20$ h, $k_{\mathrm{el}} = 0.05$ h^{-1}

$$[\mathrm{AUC}]_0^\tau = [\mathrm{AUC}]_0^\infty - [\mathrm{AUC}]_\tau^\infty = \frac{FD}{V_\mathrm{d}\,k_\mathrm{el}} - \frac{FD - Q_{\mathrm{el}(0-\tau)}}{V_\mathrm{d}\,k_\mathrm{el}} = \frac{Q_{\mathrm{el}(0-\tau)}}{V_\mathrm{d}\,k_\mathrm{el}}$$
$$= \frac{FD - Q_\tau}{V_\mathrm{d}\,k_\mathrm{el}} \tag{8.2}$$

Equation (8.2) reveals that $[\mathrm{AUC}]_0^\tau$ is proportional to FD corrected in terms of the amount of drug in the body at time τ, Q_τ; the latter quantity is not only related to absorption characteristics of the formulation, but also to drug elimination characteristics, Eq. (8.1). Thus, the area $[\mathrm{AUC}]_0^\tau$ can be used as an indicator of the extent of drug's absorption. In the same vein, Eqs. (8.3) and (8.4) were derived for two-compartment model drugs following zero- and first-order absorption, respectively.

$$[\mathrm{AUC}]_0^\tau = \frac{FD}{V_\mathrm{d}\tau}\left[\frac{k_{21}-\alpha}{\alpha^2(\beta-\alpha)}\left(\alpha\tau + e^{-\alpha\tau} - 1\right) + \frac{k_{21}-\beta}{\beta^2(\alpha-\beta)}\left(\beta\tau + e^{-\beta\tau} - 1\right)\right] \tag{8.3}$$

$$[\mathrm{AUC}]_0^\tau = \frac{k_\mathrm{a}FD}{V_\mathrm{d}}$$
$$\times\left[\frac{(k_{21}-\alpha)\left(1-e^{-\alpha\tau}\right)}{\alpha(k_\mathrm{a}-\alpha)(\beta-\alpha)} + \frac{(k_{21}-\beta)\left(1-e^{-\beta\tau}\right)}{\beta(k_\mathrm{a}-\beta)(\alpha-\beta)} + \frac{(k_{21}-k_\mathrm{a})\left(1-e^{-k_\mathrm{a}\tau}\right)}{k_\mathrm{a}(\alpha-k_\mathrm{a})(\beta-k_\mathrm{a})}\right] \tag{8.4}$$

where α and β are the distribution and elimination hybrid rate constants, k_a is the first-order absorption rate constant, τ is the duration of drug absorption, and k_{21} is the rate constant of drug transfer from the peripheral to the central compartment. Again, $[\mathrm{AUC}]_0^\tau$ is proportional to FD; in this case, the amount of drug, in the body at time τ, Q_τ will be present in the central and peripheral compartments too.

8.3 Toward the Revision of Bioequivalence Assessment

All experimental results analyzed using the PBFTPK models point to the fact that the FAT concept is physiologically sound and the estimates for τ represent a drug property for all drugs administered orally using immediate release formulations. Accordingly, the corresponding area $[\mathrm{AUC}]_0^\tau$ can be used as the metric for the assessment of the extent of absorption in bioequivalence studies. A non-compartmental methodology was developed and applied to data from a pilot bioequivalence study involving two tests (*T*1 and *T*2) and a reference (*R*) formulations of axitinib tablets. The concentration–time profiles of the mean values of the 24 subjects who participated in the study are shown in Fig. 8.5.

Initially, we fitted the PBFTPK models to the data of Fig. 8.5; the best fit results are shown in Fig. 8.6. For the three formulations studied, a model with two constant successive input rates and one compartment disposition provided the best fitting results. In all cases, this two steps axitinib absorption terminates in the time interval

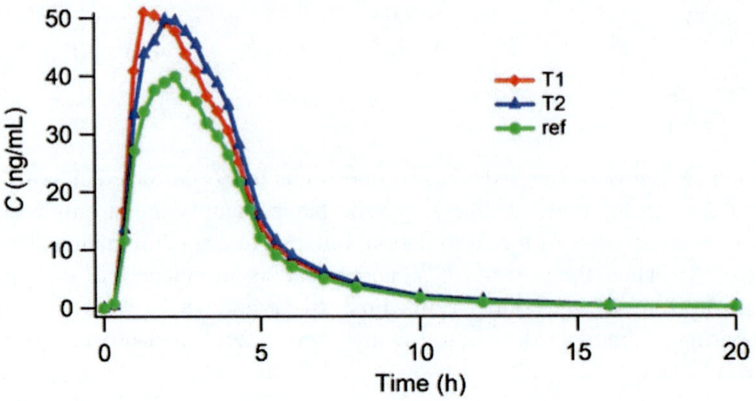

Fig. 8.5 Mean concentration–time profiles for the three axitinib formulations studied

3.0–3.3 h. It is worth mentioning that one-compartment model disposition was found in all cases studied while a two-compartment structural model with first-order absorption and lag time best described axitinib population pharmacokinetics [6]. Needless to say that PBFTPK models were not involved in Pfizer's axitinib population pharmacokinetic analysis [6].

We propose a non-compartmental approach for the assessment of bioequivalence. The methodology relies on the calculation, using the trapezoidal rule, of the ratio of the cumulative areas (*T*1/*R* and *T*2/*R*) of the blood concentration versus time curves based on the sampling points of the study [2]. These ratios exhibit a nonlinear change as a function of time reaching a plateau, Fig. 8.7. According to the FAT concept the ratio (*T*1/*R* or *T*2/*R*) reaches a plateau when the absorption processes of both the test and reference (*T*1 and *R* or *T*2 and *R*) formulations have ceased. The plateau value of this ratio is a measure of the relative bioavailability of the two formulations. For example, the estimates for *T*1/*R* and *T*2/*R* for the axitinib formulations based on the mean values are 1.22 and 1.27, respectively. Statistical criteria, e.g., 90%CI (confidence interval) can be applied using the data of all individuals of the study for the assessment of bioequivalence in terms of the extent of absorption.

8.4 Does FDA's Bioavailability Definition of 1977 Require Reconsideration?

Immediate release formulations Realizing that the Finite Absorption Time (FAT) is a physiologically sound concept (property) for all drugs administered orally in immediate release formulations, the definition of bioavailability can be altered to "*Bioavailability is the extent to which the active ingredient or active moiety is absorbed from a drug product and becomes available at the site of action.*" The rate concept should be abolished following (1) Tucker's et al. [7] suggestion "*that*

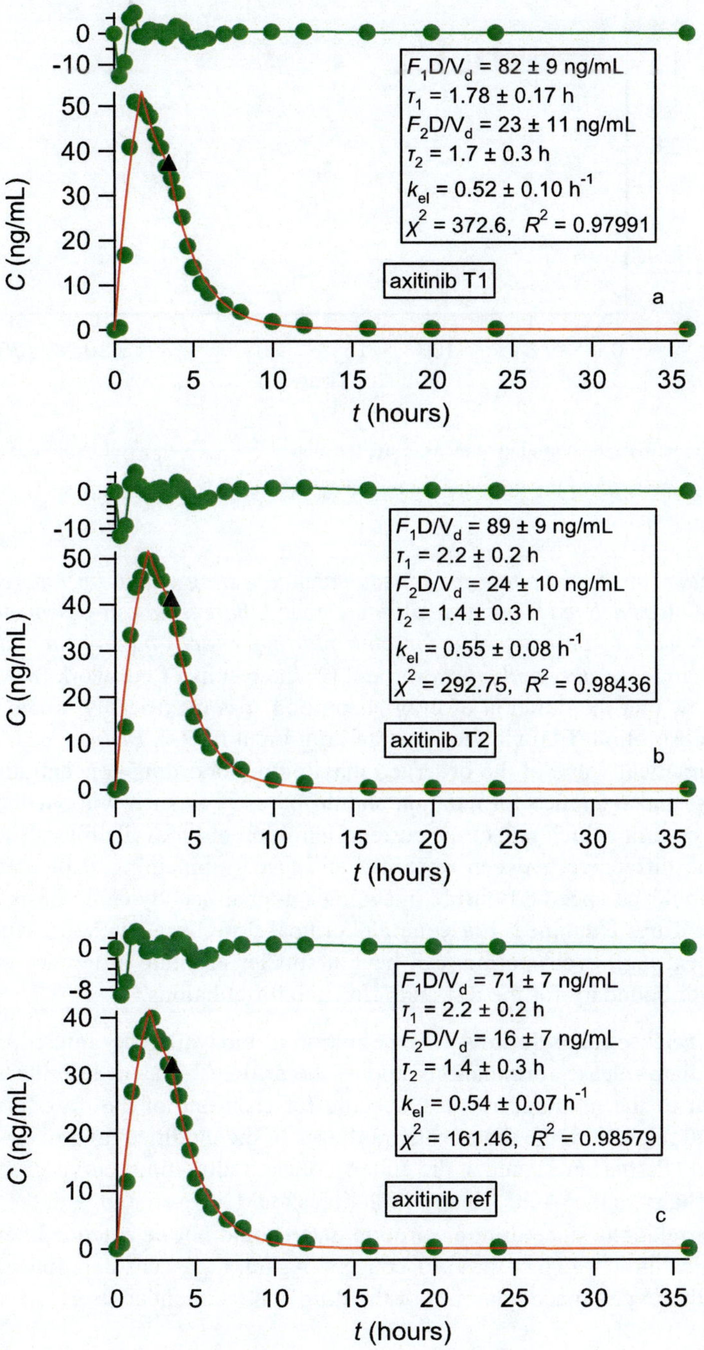

Fig. 8.6 Best fit results for the axitinib formulations *T*1, *T*2, and *R*

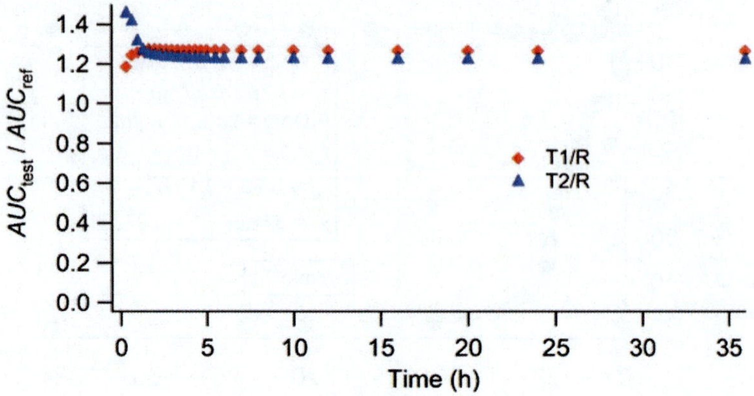

Fig. 8.7 The ratio of the cumulative areas under the curves $\frac{\sum [\mathrm{AUC}]_{t_{i-1}}^{t_i}, T}{\sum [\mathrm{AUC}]_{t_{i-1}}^{t_i}, R}$ as a function of time using the sampling points (t_i) of the study for the two axitinib tests *T*1 and *T*2 with respect to reference *R* studied

the ambiguity in the rationale for bioequivalence testing would be removed if the term "rate" was deleted from its definition" since *"there is no rate parameter which allows products to be compared for both pharmaceutical quality (actual release rate) and clinical safety and efficacy."* and (2) the results of our work [8, 9], which clearly show that the duration of drug absorption, τ is inextricably linked with the extent of absorption (FD) via the constant drug input rate(s), FD/τ.

The numerical value of the observed maximum blood drug concentration, $C_{\max}$ for the test and reference formulation should be used as such without the current statistical criteria, which reflect the current definition of bioavailability. The magnitude of the difference between reference and test formulations in bioequivalence studies should be specified on pharmacological–pharmacodynamic basis for each one of the drugs examined. For example, critical dose drugs with a narrow therapeutic index, e.g., cyclosporine, can have a smaller absolute difference and/or an upper/lower boundary for the test and reference formulations.

Modified release formulations The definition of bioavailability quoted above for the immediate release formulations applies here too. Tailor-made criteria for the assessment of the modified release, specific for each one of the drug/formulation studied and relevant to the therapeutic purposes of the modified drug release should be applied. Partial areas under the blood concentration–time curve can be used complementary to the $[\mathrm{AUC}]_0^{\tau}$ metric. Caution should be exercised with the in vitro–in vivo correlations since drug absorption data should not be extended beyond the physiological limit ~30 h of the FAT concept. Again, $C_{\max}$ considerations should be in line with the pharmacodynamics of the drug and its intended use.

References

1. Food and Drug Administration, Code of Federal Regulations Title 21, 5 Subpart B - Procedures for Determining the Bioavailability or Bioequivalence of Drug Products. 1971. https://www.accessdata.fda.gov/scripts/cdrh/cfdocs/cfcfr/ CFRSearch.cfm?fr=320.21
2. Lovering EG, McGilveray IJ, McMillan I, Tostowaryk W (1975) Comparative bioavailabilities from truncated blood level curves. J Pharm Sci 64:1521–1524. https://doi.org/10.1002/jps.2600640921
3. Tsekouras AA, Macheras P (2021) Re-examining digoxin bioavailability after half a century: time for changes in the bioavailability concepts. Pharm Res 38:1635–1638. https://doi.org/10.1007/s11095-021-03121-w
4. Sanchez N, Sheiner LB, Halkin H, Melmon KL (1973) Pharmacokinetics of digoxin: interpreting bioavailability. Br Med J 4:132. https://doi.org/10.1136/bmj.4.5885.132
5. Center for Drug Evaluation and Research, Digoxin Bioequivalency Review 76268 (2002). https://www.accessdata.fda.gov/ drugsatfda_docs/anda/2002/76268_Digoxin_Bioeqr.pdf
6. Garrett M, Poland B, Brennan M, Hee B, Pithaval YK, Amantea MA (2014) Population pharmacokinetic analysis of axitinib in healthy volunteers. Br J Clin Pharmacol 77(3): 480–492. https://doi.org/10.1111/bcp.12206
7. Tucker GT, Rostami-Hodjegan A, Jackson PR (1995) Bioequivalence-A measure of therapeutic equivalence? In: Blume H, Midha K (eds) Bio-international 2, bioavailability, bioequivalence and pharmacokinetic studies. Medpharma Scientific publishers, Stuttgart, pp 35–43
8. Chryssafidis P, Tsekouras AA, Macheras P (2021) Revising pharmacokinetics of oral drug absorption: II bioavailability-bioequivalence considerations. Pharm Res 38:1345–1356. https://doi.org/10.1007/s11095-021-03078-w
9. Chryssafidis P, Tsekouras AA, Macheras P (2022) Re-writing oral pharmacokinetics using physiologically based finite time pharmacokinetic (PBFTPK) models. Pharm Res 39:691–701. https://doi.org/10.1007/s11095-022-03230-0